Research Methods for Science

MICHAEL P. MARDER

The University of Texas at Austin

CAMBRIDGE
UNIVERSITY PRESS

CAMBRIDGE
UNIVERSITY PRESS

University Printing House, Cambridge CB2 8BS, United Kingdom

Published in the United States of America by Cambridge University Press, New York

Cambridge University Press is part of the University of Cambridge.

It furthers the University's mission by disseminating knowledge in the pursuit of education, learning and research at the highest international levels of excellence.

www.cambridge.org
Information on this title: www.cambridge.org/9780521145848

First published 2011
3rd printing 2014

Printed in the United States of America by Sheridan Inc.

A catalog record for this publication is available from the British Library

ISBN 978-0-521-14584-8 Paperback

Additional resources for this publication at www.cambridge.org/Marder

RESEARCH METHODS FOR SCIENCE

A unique introduction to the design, analysis, and presentation of scientific projects, this is an essential textbook for undergraduate majors in science and mathematics.

The textbook gives an overview of the main methods used in scientific research, including hypothesis testing, the measurement of functional relationships, and observational research. It describes important features of experimental design, such as the control of errors, instrument calibration, data analysis, laboratory safety, and the treatment of human subjects. Important concepts in statistics are discussed, focusing on standard error, the meaning of p-values, and the use of elementary statistical tests. The textbook introduces some of the main ideas in mathematical modeling, including order-of-magnitude analysis, function fitting, Fourier transforms, recursion relations, and difference approximations to differential equations. It also provides guidelines on accessing scientific literature, and preparing scientific papers and presentations. An extensive instructor's manual containing sample lessons and student papers is available at www.cambridge.org/Marder.

MICHAEL P. MARDER is Professor of Physics at The University of Texas at Austin. He is co-director and co-founder of UTeach, a program preparing secondary teachers of science and mathematics. He has been teaching a course on how to do scientific research which led to the writing of this textbook. He is author of the graduate text *Condensed Matter Physics*.

Contents

Preface

This book accompanies a one-semester undergraduate introduction to scientific research. The course was first developed at The University of Texas at Austin for students preparing to become science and mathematics teachers, and has since grown to include a broad range of undergraduates who want an introduction to research. The heart of the course is a set of scientific inquiries that each student develops independently. In years of teaching the course, the instructors have heard many questions that students naturally ask as they gather data, develop models, and interpret them. This book contains answers to those most common questions.

Because the focus is on supporting student inquiries, the text is relatively brief, and focuses on concepts such as the meaning of standard error, p-values, and deterministic modeling. If a single statistical test, such as χ^2, is adequate to deal with most student experiments, the text does not introduce alternatives, such as ANOVA, even if they are standard for professional researchers to know.

The mathematical level of the book is intermediate, and in some places presumes knowledge of calculus. It could probably be used with students who don't know calculus, skipping these sections without great loss.

There is an instructor's manual that describes daily activities for a 14-week class that meets two hours per week in a classroom and two hours per week in a lab. It is available at www.cambridge.org/Marder. The classroom sessions are not lectures covering the material in these chapters, but instead consist in activities focusing on basic concepts. The text in many cases contains more complete explanation than there is time to deliver in class.

The basic idea for the class and hence of this book is due to David Laude, Professor of Chemistry and Associate Dean for Undergraduate Education at UT Austin. He had two essential insights: First, the way to learn about scientific research is actually to do some. Second, it doesn't matter if research results are not new so long as they are new to the person who does them. The ingenious order

"BE CURIOUS!!" in the first inquiry assignment (page 15) comes from his first assignment in the first semester he taught it.

Many other course instructors have contributed. Mary Walker and Denise Ekberg both brought in course elements because of their backgrounds that span scientific research and secondary teaching. They emphasized the importance of procedures to ensure student safety, and also insisted on rubrics and checklists so that students received clear messages during an otherwise free-wheeling class of what was acceptable, what was forbidden, what was desired, and what was discouraged. Thomas Hills emphasized the importance of open questions. Many teaching assistants have also contributed to the course content, particularly Sed Keller, who wrote the first draft of the appendix on use of spreadsheets.

For many years, I have co-taught the class with Pawan Kumar, Professor of Astronomy, and Dan Bolnick, Associate Professor of Integrative Biology. Pawan Kumar helped create all the homeworks, and insisted we find ways to tie research on closed questions back into research on areas of social concern. Dan Bolnick gently prodded me to throw out all previous approaches to statistics, and make use of examples from biology that worked much better.

Finally, I would like to thank Mary Ann Rankin, Dean of the College of Natural Sciences, who insisted passionately from the start that future teachers in our UTeach program learn about scientific research, and has provided every form of support needed to help the class grow.

Preparation of this text was partially supported by the National Science Foundation under DMR – 0701373.

1

Curiosity and research

1.1 Course goals

Science starts with curiosity about the world. It starts with questions about why the sky is blue, why dogs wag their tails, whether electric power lines cause cancer, and whether God plays dice with the universe. But while curiosity is necessary for science, it is not enough. Science also depends upon logical thought, carefully planned experiments, mathematical and computer models, specialized instruments, and many other elements. In this course, you will learn some of the research methods that turn curiosity into science. In particular, you will learn to

- Create your own experiments to answer scientific questions.
- Design experiments to reduce systematic and random errors and use statistics to interpret the results.
- Use probes and computers to gather and analyze data.
- Treat human subjects in an ethical fashion.
- Apply safe laboratory procedures.
- Find and read articles in the scientific literature.
- Create mathematical models of scientific phenomena.
- Apply scientific arguments in matters of social importance.
- Write scientific papers.
- Review scientific papers.
- Give oral presentations of scientific work.

You will not just be learning about these skills. You will be acquiring and applying them by carrying out scientific inquiries.

1.2 Kinds of questions

Testable questions Every scientific inquiry answers questions, and most scientific inquiries begin with specific questions. People often wonder about questions that are hard to address, such as

- Why is the sky blue?
- What are stars and why do they move at night?
- What is sound?
- Why are leaves green?
- How do you grow the biggest possible tomatoes?
- Why do balloons pop?
- How can you cure a cold?
- Why is lighter fluid bad to drink?

Questions that begin with "Why," "What," or "How" often ask for an explanation of something that is very complex, and give little guidance about how to begin finding an answer. *Testable questions* on the other hand suggest specific activities that may provide an answer. Some examples:

- Is the sky more blue on a clear day in summer than on a clear day in winter? [Take digital photographs of the sky at noon on many days during the summer and many days during the winter with the same camera, using the same f-stop and shutter speed, and at the same location, recording temperature, humidity, and cloud cover. Compare the numerical values of the pixels in the two sets of photographs.]
- Are there chemical solvents that remove from a leaf the substances that make it green? [Grind up leaves in solvents such as water, acetone, and alcohol and check whether the green color can be extracted.]
- How does the loudness of a balloon popping depend on how much one blows it up? [Blow up balloons to different diameters and pop them at a fixed distance from a microphone connected to a digital sound recording device.]
- Is lighter fluid poisonous for ants? [Spray a mist of lighter fluid on ants, and compare their behavior with ants not sprayed with lighter fluid.]

Most of this class will be devoted to building experiments and carrying out analyses to answer testable questions. Sometimes the answers to the questions are known and one could look them up. In fact, part of the course will deal with techniques for looking up known answers. But the point of this class is not so much to answer particular questions as to provide experience in how scientists answer questions.

Closed and open questions Another way to categorize questions is to divide them between closed and open. *Closed questions* are those with a specific answer that is already known. Some examples:

- What is the name of the third planet from the Sun?
- What is the indefinite integral of $\tan \theta$ with respect to θ?
- At what rate does a ball accelerate when dropped near the surface of the Earth?
- What is the pH of an orange?
- How many vertebrae are there in the spine of a rhesus monkey?
- How large are the largest tides observed in the Bay of Fundy?

Open questions may have multiple answers, and require much more thought and interpretation. Some examples:

- How did the third planet from the Sun get there?
- Is the unproved conjecture "Every even integer greater than 2 can be represented as the sum of two primes" provable or unprovable?
- Are the fundamental constants of physics changing over time?
- Is it possible to make a completely synthetic juice that tastes as good and is as healthy as orange juice?
- Are there mutations of the rhesus monkey that are immune to any form of common cold?
- How much will the mean sea level change over the next 100 years?

Closed questions are not very challenging if someone just gives you the answer, but they can be demanding indeed if you are the one responsible for finding it. Making progress on open questions often involves answering a series of closed questions. For example, making predictions about the mean sea level over the next 100 years should be influenced by careful measurements of the mean sea level today. Checking whether the constants of nature are changing requires measuring very carefully their current values. Learning how the Earth came to its current orbit around the Sun is aided by knowing the the mass and chemical composition of the Sun and the Earth.

So, while most scientists have open questions in the backs of their minds while they work, and are motivated by large and significant issues, they spend most of their time taking small steps, answering closed questions. You should not worry if your first steps in science address apparently simple questions. Learn the process by which scientists solve simple problems well, and you will have gained skills to begin addressing the largest questions of the day.

1.3 Research methods for science

Scientists spend their time in many ways, and using a wide variety of research methods. The following sections describe some of the most common. They are based upon practical observation of how scientists spend their time and the sorts of investigations described in published papers and in conference presentations.

1.3.1 Test a hypothesis: Hypothesis-driven research

One research method in particular is usually singled out by introductory science texts and called ***the** Scientific Method*. Steps in this method are

1. State a hypothesis.
2. Design an experimental procedure to test the hypothesis.

3. Construct any necessary apparatus.
4. Perform the experiments.
5. Analyze the data from the experiment to determine how likely it is the hypothesis can be disproved.
6. Refine or correct the hypothesis and continue if necessary.

Hypothesis-driven research begins (no surprise) with a hypothesis. A *hypothesis* is a statement about the world that can be true or false, and whose truth is being tested. A valid hypothesis must be *falsifiable* (Popper, 1959). This means you should imagine actually being able to show it to be false, and you should be able to imagine evidence that would make enthusiastic supporters of the hypothesis abandon it. For example, the hypothesis that people cannot cause spoons to float in the air using mental energy could be falsified by finding a single person reproducibly capable of making spoons float without the aid of wires or other supports. When a hypothesis is *valid,* it meets the rules for being a hypothesis, whether it is true or false, while if a hypothesis is invalid it breaks the rules and should not be considered. Table 1.1 provides examples of valid and invalid hypotheses.

The U.S. National Institutes of Health, which funds most of the basic medical research in the United States, divides research into two categories: hypothesis-driven research and curiosity-driven research. The National Institutes of Health almost exclusively funds hypothesis-driven research. Both hypothesis-driven and curiosity-driven research have their place, and science needs both of them to function.

Hypothesis-driven research is most appropriate when a researcher is trying to decide between a small number of mutually exclusive cases, such as finding which of two medical treatments is more effective, or whether electric power lines cause cancer. Hypothesis testing is particularly important in medical research, since nothing is more likely to cause people to invest in ineffective procedures or drugs than hopes of a cure for themselves or loved ones. Hypothesis-driven research is designed to protect researchers against subtle biases that can distort research outcomes and are particularly difficult to avoid when life and health are at stake. Biases are also hard to avoid whenever any participants in the research have preconceived notions of what the results should be. Procedures to avoid bias in hypothesis-driven research are described in more detail in Chapter 2.

Hypothesis-driven research is not appropriate when the researcher is trying to decide between a vast number of possible cases. For example, suppose one is given a rock and wants to know its density. There is nothing to stop the researcher from beginning the investigation by saying "My hypothesis is that the density of this rock is 1 gm/cm^3" and testing the hypothesis. Starting an inquiry with a rough estimate of what the answer should be expected to be is always a good idea. However, calling

Table 1.1 Various statements that could be considered as scientific hypotheses with comments on whether they are valid or not.

Statements	Comments
Briar's Aspirin cures headaches faster than RCS Aspirin.	Valid hypothesis. Can be checked by using the two forms of aspirin on randomly chosen populations of headache sufferers.
Eating two ounces of olive oil a day decreases the odds of contracting heart disease.	Valid hypothesis. Can be checked by randomly assigning large numbers of people diets that contain or do not contain two ounces of olive oil and monitoring their health over long periods of time.
The gravitational force between two masses is proportional to the products of the masses and decreases exponentially with the distance between the masses.	Valid hypothesis. Also false, since gravitational force decreases as the square of the distance, not as the exponential.
If electrons were 10% less massive, no life would exist in the universe.	Fascinating statement, but not a valid and testable hypothesis, particularly since it is impossible to anticipate all forms that life could take.
A Toyota Camry weighs exactly 1000 kg.	This is a valid hypothesis, but it is silly if left as a hypothesis. There is no reason that the weight of a car should come out to be such a neat round number.
What is the best fertilizer to use to get large and tasty tomatoes?	Not a valid hypothesis. Hypotheses have to be definite statements, and cannot be questions.
Macs are better than PCs.	Not a valid hypothesis. It is impossible to imagine evidence that could sway the enthusiastic supporters of each kind of computer to accept the other as better.

the estimate a hypothesis is not helpful, because hypotheses are supposed to be tested rigorously and rejected if they do not meet high standards of evidence. The density of a rock can have a continuous range of values. Even if the answer must lie somewhere between 0.1 gm/cm^3 and 10 gm/cm^3 there is an infinite number of possibilities. Thus by picking a single value at the outset of the investigation and testing it, the odds are overwhelming that the hypothesis will be rejected. The problem is that it is very unlikely that anyone cares particularly whether the density of the rock is exactly 1.0 gm/cm^3 or not, and if someone were to say "I have tested the hypothesis that this rock has a density of 1.0 gm/cm^3 and rejected it" the natural response would be "Please, just tell me the density of the rock." This comment leads to other modes of research.

1.3.2 Measure a value: Experimental research (I)

1. Identify a well-defined quantity.
2. Design a procedure to measure it.
3. Perform the experiments.
4. Analyze and report on the accuracy of the results.

   Measuring a single number well is often much harder than it seems, and some of the technical issues involved are discussed in more detail in Section 2.2. Examples of measured quantities, ranging from fairly simple to very challenging, appear in Table 1.2. Some of the primary challenges in measuring values have to do with errors in the measurement process. *Random errors* reveal themselves because of continual variations in the values one obtains with repeated measurement. *Systematic errors* are more difficult to catch because the same error occurs at the same size in every measurement, and such errors can only be eliminated either by thinking through the sources of problems and removing them, or by finding completely independent ways to make the measurements and comparing them.

   Measuring a few numbers underlies many of the most expensive large group projects in science. Experimental particle physics, which employs thousands of scientists at international laboratories with budgets in the hundreds of millions of dollars, is devoted to finding masses and charges of a small number of elementary particles. The Human Genome Project (GENOME, 2008) was one of the great triumphs of science in the last 50 years, yet it consisted at a primitive level in finding a long sequence composed of four letters, which one can think of as a single very large number in base 4. Government agencies like to fund projects of this type for the simple reason that the success of the project is almost guaranteed. The researchers will come up with a collection of numbers, and those numbers are a deliverable that the government agency can display to show that the money was well spent.

   Measuring a single value is often the starting point for testing a hypothesis or measuring a series of related values. Therefore even when a single number is not the primary aim of a research project, being able to measure numbers is a basic skill that underlies all modes of research.

1.3.3 Measure a function or relationship: Experimental research (II)

1. Observe a phenomenon and develop testable questions.
2. Identify control variables and response functions.
3. Design an experimental procedure to vary the control variables, measure the response variables, and keep other factors constant.

Table 1.2 Examples of values that scientific researchers might be interested in measuring.

Value	Comments
Density of crystalline silicon at room temperature	Density cannot be obtained directly, but is defined to be the ratio of mass to volume, and so can be obtained by separately measuring mass and volume of crystalline samples and taking the ratio.
Charge of the electron	The first experimental information obtained at the end of the nineteenth century about the electron was the ratio of charge to mass. The physicist Robert Millikan won a Nobel Prize for a series of experiments between 1910 and 1920 involving oil droplets that made it possible to determine the charge separately.
Mass of the electron neutrino	The electron neutrino is one of the most plentiful particles in the universe, but it reacts very rarely with ordinary matter and is therefore almost invisible to us. All experiments to determine its mass have so far concluded that it was too small to measure, but even a very slight mass has great implications for the total amount of matter in the universe. One current experiment to measure the electron neutrino mass is being conducted at the Research Center of Karlsruhe in Germany and has a budget of 33 million euros (KATRIN, 2008).
Distance from Earth to the nearest star other than the Sun.	Proxima Centauri is 4.22 light years or 39,900,000,000,000 km from Earth (NASA, 2008).
The number of base pairs and distinct genes in the human genome	The number of base pairs is around 3 billion and the number of distinct genes is around 30,000. These values were determined as part of the Human Genome Project (GENOME, 2008), a multibillion dollar scientific effort that involved government and university scientists, as well as a corporation, and provided the first sequencing of the human genome.

4. Perform the experiments.
5. Analyze the relation between control variables and response variables, and characterize the relation mathematically.

Much of experimental physics and chemistry operates according to this research method; examples appear in Table 1.3. The starting point is to identify *control variables* and *response variables*. A control variable is a quantity that the experimenter varies at will to change the character of the experiment, while a response

Table 1.3 Examples of functions or relationships that researchers might measure.

Function or relationship	Comments
Find how the speed of sound in air at fixed pressure depends upon air temperature.	The control variable is temperature, and the response variable is sound speed. This sort of experiment is extremely common in experimental physics and chemistry, and reference volumes are full of the results. Examples are provided by the hundreds of volumes called *Numerical Data and Functional Relationships in Science and Technology* by Landolt and Börnstein, published in multiple series over many decades by Springer-Verlag, originally in German, and now in English.
Place a fluid between two cylinders, rotate the outer cylinder, and find the state of the fluid as a function of the rotation speed.	The control variable is rotation speed of the cylinder. The fluid undergoes a number of abrupt qualitative transitions, from smooth uniform motion, to rolls that look like a barber pole, to turbulence.
Measure the stiffness of a sample of rubber as a function of the amount of cross-linking agent used to process it.	The control variable is cross-linking agent (Charles Goodyear used sulfur to stiffen natural rubber) and the response variable is stiffness. This type of measurement is commonplace in engineering and industrial research.
Find the average size of a specific breed of tomatoes grown in specific soil and climate as a function of the amount of salt in the soil.	The control variable is salt concentration, and the response variable is tomato size. An experiment like this that really sought to optimize the size of tomatoes would not typically focus on a single variable. Instead a series of factors such as salt concentration, fertilizer, irrigation, and seed type would all become part of an experimental designed to improve tomatoes.
Measure how often allergy sufferers sneeze per day as a function of the dose of anti-histamine they take.	The control variable is the dose of anti-histamine. Different allergy sufferers are sensitive to different substances, all of which are likely to be varying beyond control of the experimenter.

variable is some other quantity measured as an outcome. In a good experiment, all variables other than the control variables that affect the outcome are held constant. So in measurements of the speed of sound in air with respect to temperature, the pressure and humidity of air need to be kept constant. In measuring the stiffness of rubber with respect to a cross-linking agent, the temperature needs to be held constant. An experiment on the effectiveness of an anti-histamine to prevent sneezing would raise the greatest challenges of this sort, since one person might be allergic to cats, another to mold in the air, and it would be difficult either to find lots of subjects all allergic to exactly the same thing or to control the precise amount to which they were exposed.

There are many different skills involved in actually carrying out experiments, ranging from construction of apparatus and safe laboratory practice to the safe and ethical treatment of human subjects. These and other technical issues are discussed at greater length in Chapter 2.

The mathematical analysis of a largely experimental project may involve nothing more than careful characterization of error bounds associated with each measured point. Or it may involve careful comparison with a particular mathematical theory of the experiment. Sometimes the analysis involves actual construction of a mathematical model, as in the next method of research.

1.3.4 Construct a model: Theoretical sciences and applied mathematics

1. Choose a relationship discovered through experimental investigation.
2. Construct mental pictures to explain the relationship, and develop hypotheses about origins of the phenomenon.
3. Identify basic mathematical equations from which the relation might result.
4. Using analytical or numerical techniques, determine whether the experimental relationship results from the basic mathematical equations.
5. If incorrect, find a new mathematical starting point.
6. If correct, predict new relationships to be found in future experiments.

This mode of research describes much of applied mathematics, theoretical physics, theoretical chemistry, theoretical geology, theoretical astronomy, or theoretical biology (Table 1.4). For example, the experimental observation might be intense bursts of gamma-rays. A hypothesis might be that they emerge from gravitational collapse of certain stars. A lengthy process of modeling the collapse of stars, trying to calculate the radiation that emerges from them, would be needed to check the hypothesis.

In variants of this mode of research, the modeling takes place without any experimental input, and emerges with experimental predictions. In other variants, this type of research can lead to new results in pure mathematics.

1.3.5 Observational and exploratory research

1. Create an instrument or method for making observations that have not been made before.
2. Carry out observations, recording as much detail as possible, searching for unexpected objects or relationships.
3. Present results and stimulate further research.

Laboratory experiments have control variables, but in a huge variety of scientific investigations, scientists measure quantities they cannot control. This mode

Table 1.4 Examples of models of scientific phenomena.

Model	Comments
Calculate the distance a projectile fired from a barrel near the Earth's surface at known speed will travel as a function of angle.	The calculation can be performed using Newton's laws of motion. It is fairly easy if one neglects things such as air resistance and the Earth's rotation, and more challenging if these are included.
Consider a fluid placed between two cylinders with the outer one rotating and calculate the rotation speed at which steady fluid motion becomes unstable to the formation of rolls.	This computation was first carried out by Taylor (1923) using equations for fluids called the Navier–Stokes equations, and constituted the first nontrivial comparison of theory and experiment for fluid motion.
Find the arrangement of atoms in deoxyribonucleic acid (DNA).	Watson and Crick (1953) determined the structure of DNA. Although their article contains a few numerical values, the paper mainly contains the concept of DNA as a double helix with a schematic diagram, no complicated calculations.
Find the weather in the United States two days from now.	Weather prediction is one of the most computer-intensive activities in the world. The process of prediction begins with a vast collection of data from weather monitoring stations around the globe and continues with computations based upon equations for the motion of air including temperature, pressure, and humidity.
Find how the populations of animals change from year to year in an environment of fixed size and limited resources.	Simple iterative equations written down in the 1970s to describe the time development of populations led to the mathematical theory of chaos.

of research covers an enormous range of possibilities. It describes the expeditions that revealed the different continents to European explorers and mapped the globe. It describes first investigations when new scientific tools are developed. It describes the increasingly accurate maps of the night sky created by new generations of telescopes, or unexpected new particles discovered in particle accelerators. It describes much geological and biological field work. More examples are in Table 1.5.

Many research projects contain an exploratory phase, which produces something of interest that then becomes the subject of other research methods. For example, the first observation of gamma-ray bursts by Klebesedal *et al.* (1973) simply reported observations of enormously powerful far-away explosions made over several years, ruled out the possibility that they were due to known sorts of

Table 1.5 Examples of observations and explorations.

Observation	Comments
Find how the position of Mars in the sky varies over times of minutes, days, hours, and years.	Careful measurements of locations of stars go back to the beginning of recorded human history. The measurements of Tycho Brahe in the late 1500s as interpreted by Kepler played a critical role in the development of modern astronomy, physics, and mathematics.
Travel to the Galapagos islands and make careful observations about wildlife.	Charles Darwin's observations of wildlife on these islands during the second voyage of the Beagle, 1831–1836, particularly of finches whose beaks were adapted to different food supplies in different regions, played an important role in the thinking that eventually led to the theory of evolution.
Find the percentage of adult U.S. residents whose height lies between 30 and 31 cm, 31 and 32 cm... 100 and 101 cm ... 250 and 251 cm ...	This is a histogram or distribution function. It provides information that goes beyond simply recording average values. Many quantities, such as people's heights, are intrinsically variable, and distribution functions capture the full story.
Search for Soviet nuclear explosions in outer space.	The USSR never exploded nuclear weapons in outer space. However, the satellites the U.S. put in orbit to look for them detected massive explosions at vast distances from Earth called gamma-ray bursts.
Measure the chemical composition of Greenland ice cores as a function of depth.	Greenland ice cores provide annual information on climate and chemicals in the Earth's atmosphere going back hundreds of thousands of years. One international project was NGRIP (2003), which obtained a 3-km-long core.
Use fMRI to map regions of activity in the brain of an individual undergoing an epileptic seizure.	Functional Magnetic Resonance Imaging (fMRI) is a tool with the ability to provide three-dimensional images of the interior of the human body, and it produces the images rapidly enough that it can even provide information on the state of the brain during specific activities.

exploding stars, and stopped. The observation triggered a long-term research effort with both theoretical and further observational components that continues to this day. The National Aeronautics and Space Administration (NASA) launched a dedicated satellite in 2004 (SWIFT) to observe gamma-ray bursts.

Many U.S. government agencies are devoted to gathering data. Even if the techniques used to assemble the data are routine and would not be called forefront research, the methods used to gather the data are scientific. The National Institute of Standards and Technology (formerly the National Bureau of Standards) is responsible for maintaining weights and measures as required by the U.S. Constitution. Other agencies such as the Centers for Disease Control gather data about the U.S. population. Acquiring such data requires safe treatment of human subjects and methods to arrive at conclusions about a large population by sampling portions of it, techniques that will be described in Section 2.4.

1.3.6 Improve a product or process: Industrial and applied research

1. Identify market need for product.
2. Design product with the potential to meet the need.
3. Build prototype products.
4. Determine whether products function as desired.
5. Optimize products with respect to cost, speed, environmental consequences, and other factors that affect profit.
6. Bring product to market and continue.

The importance of science, scientists, and engineers in developing products has risen and fallen over the years. In the nineteenth century, national economies were based more on trades and traditional practices than on scientific methods. During the twentieth century, almost every industry and form of human activity was transformed by science and its offshoots. To mention just one example, scientific methods in farming meant that people no longer had to grow their own food, and most of the U.S. population moved to urban and suburban areas. Other examples appear in Table 1.6. Companies became so enamored of the benefits of scientific research that many established large facilities with a basic research mission. In its heyday in the 1960s, Bell Telephone Laboratories even had a composer of modern music on staff. Most of these large industrial basic research laboratories have now closed because the time interval between a basic scientific discovery and its use to make a profit is too long. Still, most companies have one or more divisions devoted to Research and Development. The research is closely tied to development of new products, and is more likely to employ people with training in engineering than in pure science. The scale of industrial research is indicated by the fact that in 2008 U.S. research and development spending was $398 billion; the business sector spent $268 billion, mainly on product development, while the federal government spent $104 billion, mainly on research (National Science Board, 2010). Some industrial research is published in scientific journals, some is patented, and some is held tightly secret.

Table 1.6 Examples of products related to industrial research.

Product	Comments
Create a device to make light from electrical current.	Finding a long-lasting filament for an electric light bulb from carbonized bamboo was one of the great accomplishments of Thomas Edison's research team. A picture of a light bulb is a symbol for the idea of discovery, but real the story of the invention is complicated (Friedel and Israel, 1987). Edison ends up with credit for the bulb in part because he made it commerically successful and simultaneously created an electric power company.
Create a device from semiconducting materials capable of amplifying electrical currents.	The transistor resulted from an intensive three-year research effort at the Bell Telephone Laboratories in the late 1940s, led by William Shockley and employing John Bardeen, who is the only person ever to have won two independent Nobel Prizes in physics. This invention was the seed from which the entire electronics industry soon sprouted, and it inspired many large companies to fund basic research laboratories.
Create new small-molecule inhibitors of protein Kinase B for use as anti-cancer agents.	This example comes from a 2005 business agreement between Astra-Zeneca and a smaller firm Astex, and illustrates the very specific and goal-oriented nature of drug design. Developing a new drug involves choosing a disease whose sufferers can afford to pay for a cure, finding a cure, demonstrating its effectiveness and establishing the seriousness of side effects with extensive human trials, and marketing the drug to patients and doctors (Astra-Zeneca).
Design an airplane for trans-Atlantic flights without any wind tunnel testing.	The Boeing 777 was designed completely on computers, mainly using a program called CATIA first developed by Dassault in France (CATIA). Design on the computer made it possible to ensure that all the pieces of the airplane fit together before any of them were actually built.
Create inherently tacky elastomeric, solvent-dispersible, solvent-insoluble, acrylate copolymer microspheres.	This is the technical description of the invention that made Post-It Notes possible. It was patented in 1970 by Spencer Silver, a chemist working for 3M corporation, and after a challenging marketing campaign, because the product was initially so unfamiliar, Post-Its were distributed across the U.S. by 1980. The essential idea is to have an adhesive strong enough to hold paper, but weak enough to release the paper without tearing, and that can stick multiple times.
Write programs that regulate and control scientific instruments through computers rather than with physical knobs and switches.	Labview was first released in 1986 by National Instruments after three years of research and development. It is a complete programming language that makes it possible to control scientific instrumentation from a circuit diagram the user draws on the screen, and is now standard equipment in experimental laboratories.

1.3.7 Allied areas of research

There are many additional skills and areas of research that make scientific research possible. Here are three:

Library research *There is nothing like a few months in the lab to save a few hours in the library.* – ANON.

Because of the vast quantity of research that has already been performed, it is irresponsible to move very far through a project without attempting to determine whether the answer is already known. It is also difficult to determine with certainty whether a problem one is working on has been solved. If the problem has been solved, then the fastest way to find an answer to a scientific question is to look up the answer. Reading scientific papers can be even harder than finding them because of the specialized knowledge so often needed to make sense of them. Chapter 5 will discuss in more detail how to find results in the scientific literature, and how to read a scientific paper.

Pure mathematics *An engineer thinks that his equations are an approximation to reality. A physicist thinks reality is an approximation to his equations. A mathematician doesn't care.* – ANON.

Most of science cannot be practiced without mathematics, and much of pure and abstract mathematics was developed because of the desire to solve scientific problems. However research in pure mathematics is sufficiently different from scientific research that it is listed as an allied area rather than as part of science. Developing pure mathematics means developing conjectures about relations that might be true, defining new entities, and proving theorems. It is extraordinary that such an elaborate body of knowledge has developed from nothing but pure thought. Statements in science are accepted because they have been checked many times and (almost) always come out to be true, but mathematics has a much higher standard for truth. Mathematical statements have to be proven and once proven they provide the most certain knowledge people have.

Computers and computer science *Computer science is no more about computers than astronomy is about telescopes.* – EDSGER WYBE DIJKSTRA

Computers have changed the way science is practiced in many ways. Data are gathered electronically, scientific papers are composed and distributed electronically, and computational science is arguably a branch of science distinct both from experiment and theory. Familiar programs such as Excel bring into millions of households computational possibilities that were available only at military laboratories 50 years ago, and almost unimaginable 50 years before that.

Several different disciplines have contributed to making this revolution possible. Improving the hardware that lets computers continue to become faster involves many different fields, but mainly electrical and electronic engineering. Thinking of new ways to organize computers and new ways to program them is the domain of computer science. Computer science has its roots in formal logic and discrete mathematics, but has now become a separate discipline with its own goals and standards.

1.4 Putting methods together

In any given scientific research project, one of these research methods may best describe the project overall, but many of the others will often come into play as essential skills. For example, to test the hypothesis that hairspray affects the color of fabrics (Test a Hypothesis) , one will need careful quantitative measurements of the color of fabrics (Measure a Value). To measure the charge on the electron through the motion of oil drops (Measure a Value), one may need to test the hypothesis that the viscosity of air does not depend significantly on temperature (Test a Hypothesis). Having determined that viscosity of air does indeed depend upon temperature, one may next need to find the form of the dependence (Measure a Function or Relationship). Elements of each research method enter into almost every substantial scientific project.

The idea that scientific research takes so many forms may be unwelcome to those who would like a single simple rubric to apply to all scientific investigations. The philosopher of science Feyerabend (1975) makes the case that there is no method to science at all – "Anything Goes!" – but he goes too far. Just because music includes more than just pieces for symphony orchestra does not mean that all sounds are music. The research methods described in this chapter provide an overview of methods used by practicing scientists today. There is a fair amount of freedom in how to proceed, but there also are many standards applied to each piece of work before it can be accepted by other scientists. The following chapters will lay out in more detail the techniques and tools scientists use in performing their research and judging the work of others.

Assignments

1.1 **Home inquiry**

    Background For this assignment, you need to perform a bit of scientific inquiry away from an academic setting. Write a paper that describes what you learned. It is up to you to decide what you will investigate, as well as the format for the write-up. The results do not have to be terribly long.

Selecting a topic Here is the hard part. BE CURIOUS!! You do not need anything more than what can readily be accomplished around the house, dorm, or outdoors. Look around you and let your curiosity take hold. However, **do not carry out any project involving human subjects or vertebrate animals** unless you verify with your instructor that there are no safety concerns with your idea.

Length 1–2 typed pages.

Yes, you are being asked to do an assignment before you have learned anything in this course. This effort will give you an opportunity to find out just how much you already know, or don't know, about science in action. If you take a look at the course outline, you will be learning about experimental design, statistics, modeling, and information as the semester goes along. For now, though, please don't do anything other than what you already know how to do. You can use this assignment for comparison with assignments you produce after you learn more. In other words, this write-up may not be very good, but that's OK.

1.2 Inquiry grading

Background Your instructor will supply you with a report on a scientific project to read. You should read it, and evaluate it according to the Inquiry Grading Rubric in Appendix D. A checklist for you to fill out is available on the course website.

What to turn in You should fill out the checklist assigning points to the project you have read, and turn in the checklist. However, even more important is that you write comments on the report that you could give to the author to help him improve it. You should type up these comments and submit them. They should be a minimum of half a page. If you want to make additional written comments on your copy of the project and turn that in as well, you may.

References

Astra-Zeneca, Search for "astex" at http://www.astrazeneca.com, Retrieved March 2010.

CATIA (Retrieved March 2010), http://en.wikipedia.org/wiki/catia

P. K. Feyerabend (1975), *Against Method, Outline of an Anarchistic Theory of Knowledge*, Verso, UK.

R. Friedel and P. Israel (1987), *Edison's Electric Light: Biography of an Invention*, Rutgers University Press., New Brunswick, New Jersey.

GENOME (2008), http://www.genome.gov/11006943

KATRIN (2008), http://www-ik.fzk.de/~katrin/

R. W. Klebesedal, I. B. Strong, and R. A. Olson (1973), Observations of gamma-ray bursts of cosmic origin, *Astrophysical Journal*, **182**, L85.

NASA (2008), http://heasarc.gsfc.nasa.gov/docs/cosmic/nearest_star_info.html

National Science Board (2010), Science and Engineering Indicators 2010, http://www.nsf.gov/statistics/seind10

NGRIP (2003), http://www.gfy.ku.dk/~www-glac/ngrip/hovedside_eng.htm

K. Popper (1959), *The Logic of Scientific Discovery*, Routledge, London.

SWIFT (2007), http://swift.gsfc.nasa.gov

G. I. Taylor (1923), Stability of a viscous liquid contained between two rotating cylinders, *Philosophical Transactions of the Royal Society of London. Series A*, **223**, 289–343. http://www.jstor.org/stable/91148

J. D. Watson and F. H. C. Crick (1953), Molecular structure of nucleic acids, *Nature*, **171**, 737.

2

Overview of experimental analysis and design

2.1 Hypothesis-driven experiments

2.1.1 Null and alternative hypotheses

The basic outlines of hypothesis-driven research were provided in Section 1.3.1, but when setting out to create an experiment to test a hypothesis, there is much more to consider.

Your experiment should have at least one control or *independent variable*, and at least one response or *dependent* variable. The independent variable is something you are sure will change during the course of experimental measurements. The dependent variable is what you will measure, although you may not be sure if it will change. Indeed the point of the experiment may be to determine if it does or does not change. In laboratory experiments it is common to call the independent variable a control variable because you decide upon its value and control it. For example, in an experiment concerning the effect of fertilizer doses on size of tomatoes, you control the concentration of fertilizer given to each plant. In observational research, the independent variable may be something you cannot control. For example, you might conduct a study of tomatoes grown outside and investigate whether they grow faster on hot days than on cool days. In this case, nature will bring you the hot and cold days; you cannot control when they will happen, but you can still study their effects. Experiments are usually better when the investigator controls the independent variable, but this is not always possible.

Another way to think about the relation between the independent variable and the dependent variable is in terms of cause and effect. The independent variable is something you believe may cause a change in the dependent variable. Finding causes is important because it lets one make predictions and gives a guide for action. If Briar's Aspirin really does cure headaches faster then RCS, then pharmacists should recommend it. If airplanes really are more fuel-efficient with winglets,

18

Table 2.1 Examples of alternative hypotheses and null hypotheses.

Alternative hypothesis	Null hypothesis
Briar's Aspirin cures headaches faster than RCS Aspirin.	Briar's Aspirin and RCS Aspirin cure headaches at the same rate.
Eating two ounces of olive oil a day decreases the odds of contracting heart disease.	Eating two ounces of olive oil a day has no measurable effect on the odds of contracting heart disease.
A Toyota Camry weighs exactly 1000 kg.	A Toyota Camry does not weigh exactly 1000 kg.
Tall college students participate in intramural sports more than short college students.	Height does not affect the rate at which college students participate in intramural sports.
Mac users like their computers more than PC users do	Users of Macs and PCs like their computers equally well.
Airplanes are more fuel efficient when they are equipped with curved winglets at the end of each wing.	Winglets have no effect on the fuel efficiency of airplanes.
The gravitational force between two masses is proportional to the products of the masses but decreases as the fifth power of the distance between the masses for distances much less than 10^{-6} m.	The gravitational force between two masses is proportional to the products of the masses and decreases as the square of the distance between the masses for all distances. [The natural null hypothesis here is not that two masses will not attract each other at all, but that the standard laws of physics hold.]

and if the winglets' cost is less than the cost of the extra fuel consumed over time, airlines will install them.

Once you have settled on an independent variable, you can state a *null hypothesis*. The null hypothesis is usually that changes in your independent variable will not lead to any significant changes in your dependent variable. The hypothesis that changing the first variable does change the second is called the *alternative hypothesis*. Examples of null and alternative hypotheses appear in Table 2.1. The final example in the table shows a different form of null and alternative hypothesis where the null hypothesis is an established physical law, and the alternative hypothesis is a proposed alteration of established physical laws.

2.1.2 The problems of error, time, and money

A number of problems must be solved in every experiment. One of them is that all measurements have error. A second is that time and resources always limit the number of measurements that can actually be performed. These two problems are

Table 2.2 Measurements of time for a ball to descend
a channel with and without lubricant.

Descent time in unlubricated channel (seconds)		Descent time in lubricated channel (seconds)	
0.303	0.291	0.352	0.317
0.301	0.314	0.331	0.332
0.292	0.308	0.327	0.340
0.296	0.298	0.315	0.324
0.291	0.336	0.313	0.285

in direct conflict with each other. The effects of random error can be eliminated
by taking a vast number of measurements, but that takes a vast amount of time
and money. Time and cost can be driven down by taking a minimal number of
measurements, but the results can be plagued with error.

2.1.3 Three examples

Three examples may help illustrate how these problems arise in practice.

Ball on a ramp

An experimenter lets a ball roll down a metal channel 50 cm in length and measures
the time needed to go from top to bottom. The time is measured by having the ball
pass through a first optical gate at the top of the channel which starts a clock and
through a second optical gate at the bottom which stops it. After 10 measurements
are taken, the channel is then sprayed with a thin layer of lubricant, and the time
needed to roll the ball down the channel is measured 10 times again. The null
hypothesis is that the lubricant has no effect on the time it takes to roll down the
channel. The alternative hypothesis is that the time will become measurably greater
or smaller. The effect could go either way because the lubricant might speed up the
ball by lubricating it, or on the other hand might slow it down by being a little
sticky. Data from this experiment are shown in Table 2.2.

The data are scattered. If the lubricant had an effect, it was comparable to the
difference from one trial to the next in rolling a ball down the channel in supposedly
exactly the same way. *Did lubricant affect the time the ball needs to roll or not?*

Fish in two lakes

A researcher travels to Canada and visits two lakes, Dugout and Ormond. Small
fish called three-spine sticklebacks live in each lake. Ormond is the larger of the

two and has trout, which are a predator for the stickleback. Dugout has no trout or other predators. Three-spine sticklebacks have (astonishing as it may seem) three spines on their backs, which deter trout from eating them. One would think that the larger the stickleback, the more effective the spines in warding off trout. Therefore the spines on the sticklebacks in Ormond should be larger than in Dugout. This is the alternative hypothesis, while the null hypothesis is that spines on the sticklebacks in the two lakes are of the same size. The researcher spends two weeks collecting fish from the two lakes and measuring their spines. At the end of this time she has the measurements in Table 2.3. The fish spines have a wide range of lengths and it is far from clear through simple inspection that there is a difference between the two lakes. Even worse, there must be tens of thousands of fish in the lakes, and one has only measured around 200 of them, which is only a tiny fraction of the whole. *Are the fish spines in Ormond really bigger than those in Dugout?*

Flipping a coin

Someone hands a researcher a coin and says "I flipped this coin and I don't think it's fair. Heads and tails don't come up equally often." The researcher decides to check. He flips it 100 times and gets the results in Table 2.4. There are 54 tails and 46 heads, not exactly 50–50. *Is the coin fair?*

2.1.4 Solutions with unlimited time and money

For each of these experiments, there are solutions that would involve huge expenditures of resources. For the ball rolling down a ramp, one could build the ramp out of optical glass, flat to less than a micron, place it in a windless temperature-controlled chamber on a vibration isolation table, launch the ball with a gate run by a computer-controlled motor, and measure the passage of the ball through the optical beams with analog-to-digital cards connected to computers capable of resolving time differences of a microsecond. The accuracy of the experiment would increase by around 1000 times, for a budget of around $100,000. In the case of the fish, one could build dams and pumping stations, completely drain the lakes, and hire several dozen workers to collect all the (dead) fish in the lakes and measure their spines. Then one would know exactly the spine length of every last fish in both lakes and would not have to worry about only measuring a small fraction of them. The cost here would probably be around $500,000, not counting the environmental lawsuits. Finally, in the case of the coin, one could flip it a billion times. That ought to be enough. At two seconds per flip, it would take a little over 63 years, flipping night and day without sleep. This experiment is cheap but it does take a long time.

Table 2.3 Fish spine lengths in millimeters collected from two
Canadian lakes.

Fish spine lengths from Ormond			Fish spine lengths from Dugout			
4.83	5.01	5.21	3.82	4.54	4.79	5.06
4.58	6.21	5.9	5.44	4.4	4.93	4.51
4.35	5.95	4.71	4.59	4.64	4.67	3.85
4.12	4.18	4.14	5	4.27	4.03	4.85
5.4	4.37	5.14	5.98	4.85	4.86	4.85
4.43	5.09	4.89	4.3	4.32	5.25	4.59
4.15	4.67	5.74	4	4.71	4.78	5.12
4.78	5.21	5.29	5.07	5.05	4.98	5.13
3.92	4.87	5.35	4.92	3.92	4.95	5.45
5.66	5.15	6.13	4.57	5.19	3.74	4.93
5.14	4.52	5.87	3.32	3.43	4.53	5.23
4.63	5.36	4.8	4.88	5.57	5.42	3.9
6.48	4.59	5.38	4.52	3.97	4.91	4.5
3.94	4.09	3.28	5.53	5.39	3.8	4.17
6.48	5.67	5	4.58	5.06	5.59	4.6
4.84	4.57	5.05	4.43	5.19	4.54	4.85
5.76	4.95	3.81	5.15	4.83	4.54	4.72
5.65	4.15	4.39	4.66	5.55	5.07	4.07
3.63	5.78	3.51	3.53	4.51	5.19	3.97
5.15	4.56	4.2	5.15	4.25	4.36	3.63
5.5	6.36	3.86	4.1	4.67	5.1	3.55
6.02	4.71	4.63	5.43	4.08	4.3	4.76
5.31	4.52	4.1	4.53	5.17	4.11	4.4
5.11	4.98	3.58	5.33	6.2	5.45	3.7
5.13	4.85	3.78	5.75	4.7	4.59	3.88
6.48	5.96	4.14	5.03	5.03	5.42	4.83
5.47	4.62	3.96	5	5.65	4.92	4.23
6.01	5.27	4.14	4.44	4.24	4.3	2.83
4.77	6.14	4.16	4.67	4.62	4.92	4.93
5.37	4.93	4.12	4.47	4.75	4.77	4.24
4.89	5.31	5.09	4.18	4.26	5.54	4.75
3.7	4.58	4.76	4.71	4.62	4.58	4.52
5.17	3.91	4.91				
6.17	4.54	5.49				
6.47	4.07	5.34				

2.1.5 Solutions with error analysis

The solutions involving unlimited time and money are not meant to be taken seriously, although some projects are so important that resources of that order can be devoted to them (for example, building and launching the Hubble Space Telescope, or construction of the Panama Canal). But science would not get very far if answering every small question meant mobilizing national resources. Instead,

Table 2.4 Results of flipping a coin 100 times.
1 represents heads and 0 represents tails.

0	0	1	0	0	0	1	0	1	1
0	0	0	1	1	1	0	0	0	1
1	0	1	1	0	0	1	0	1	1
1	1	1	0	0	1	1	1	0	0
0	0	1	1	0	0	1	1	0	1
0	0	0	0	0	0	1	1	0	0
0	0	0	1	1	0	1	0	1	0
1	0	1	0	0	1	1	1	0	1
1	1	0	0	0	0	0	1	1	1
0	0	0	0	0	1	1	1	1	0

scientists analyze the errors that arise during experiments and make use of information from statistics to decide whether results are significant or not. The concepts and techniques needed will be explained in more detail in Chapter 3. Here is a compact outline of the basic process:

1. For each quantity x that has been measured, compute the *sample average* $\bar{x}$.
2. For each quantity that has been measured, compute the *sample standard deviation* s.
3. For each quantity x that has been measured (N times), compute the *standard error of the mean* $\Delta x = s/\sqrt{N}$.
4. Graph each measured quantity, putting a point at the sample average $\bar{x}$, and surrounding it with error bars whose size is given by the standard error, $\Delta x = s/\sqrt{N}$.
5. If the error bars of two different quantities are very far from overlapping, the null hypothesis that they are the same can be rejected. If the error bars of two different quantities overlap a great deal, the null hypothesis that they are the same cannot be rejected.
6. For greater certainty in all cases, and particularly when error bars nearly or slightly overlap, employ more sophisticated statistical tests such as t tests or χ^2 tests to determine the likelihood that observed results could have arisen by chance alone.

Notation for standard error There are several different possible notations for the standard error. Here, you will see the standard error of the sample mean $\bar{x}$ referred to as Δx, which you can think of as the uncertainty in $\bar{x}$. Another common notation for exactly the same thing is $\text{SE}(\bar{x})$. Yet another is $s_{\bar{x}}$. In chemistry Δx usually refers to the differences of two quantities, x_1 and x_2, not the standard error.

In any given paper or book, notation should be consistent, but when you go from one paper or book or discipline to another, familiar symbols may mean something new, and familiar ideas may be denoted in different ways.

2.1.6 Application to the three examples

For each of the three examples, ball on a ramp, fish in two lakes, and flipping a coin, here is how error analysis would look.

Ball on a ramp

The time it takes a ball to roll down a channel is conceptually very well defined. The problem in measuring it comes mainly from variations in how the ball is released, unevenness in the surface of the ball and of the channel where it rolls, and limitations in the reliability of the electronic circuit that responds to the ball reaching the end point of the experiment. These errors should be expected to vary in sign and magnitude from one trial of the experiment to another, and can plausibly be eliminated by averaging.

1. The average time needed for the ball to roll down in the unlubricated channel is given by Equation (3.1) as

$$\bar{t}_{unlub} = \frac{\begin{array}{c} 0.303 + 0.301 + 0.292 + 0.296 + 0.304 \\ + \quad 0.291 + 0.314 + 0.308 + 0.298 + 0.336 \end{array}}{10} = 0.304 \qquad (2.1)$$

while for the lubricated channel one has

$$\bar{t}_{lub} = \frac{\begin{array}{c} 0.352 + 0.331 + 0.327 + 0.315 + 0.313 \\ + \quad 0.317 + 0.332 + 0.340 + 0.324 + 0.285 \end{array}}{10} = 0.324. \qquad (2.2)$$

2. Next, compute the sample standard deviations for each set of measurements. The formula is given by Equation (3.7) and is in this case

$$s_{unlub} = \sqrt{\frac{\begin{array}{c}(0.303 - 0.304)^2 + (0.301 - 0.304)^2 \\ +(0.292 - 0.304) + (0.296 - 0.304)^2 \\ +(0.304 - 0.304)^2 + (0.291 - 0.304)^2 \\ +(0.314 - 0.304)^2 + (0.308 - 0.304)^2 \\ +(0.298 - 0.304)^2 + (0.336 - 0.304)^2 \end{array}}{10 - 1}} = 0.013 \qquad (2.3)$$

$$s_{lub} = \sqrt{\frac{\begin{array}{c}(0.352 - 0.324)^2 + (0.331 - 0.324)^2 \\ +(0.327 - 0.324)^2 + (0.315 - 0.324)^2 \\ +(0.313 - 0.324)^2 + (0.317 - 0.324)^2 \\ +(0.332 - 0.324)^2 + (0.340 - 0.324)^2 \\ +(0.324 - 0.324)^2 + (0.285 - 0.324)^2 \end{array}}{10 - 1}} = 0.018. \qquad (2.4)$$

This means that for the first set of measurements, the typical amount by which individual values differ from the average is 0.013, while for the second set of measurements the typical difference between individual measurements and the average is 0.018.

3. Next, compute the standard errors:

$$\frac{s_{\text{unlubricated}}}{\sqrt{N}} = \frac{0.013}{\sqrt{10}} \approx 0.004 \tag{2.5}$$

$$\frac{s_{\text{lubricated}}}{\sqrt{N}} = \frac{0.018}{\sqrt{10}} \approx 0.006. \tag{2.6}$$

The uncertainty in the mean time for the ball to roll down the unlubricated channel is 0.004 sec and the uncertainty in the mean time for the ball to roll down the lubricated channel is 0.006. One combines the sample mean and standard error to write the estimates

$$t_{\text{unlubricated}} = 0.305 \pm 0.004 \text{ sec} \tag{2.7}$$

$$t_{\text{lubricated}} = 0.324 \pm 0.006 \text{ sec.} \tag{2.8}$$

4. Graphically, the results can be presented as in Figure 2.1. Think of the error bars as showing the range where the true mean is most likely to fall.
5. The top of the error bar for the ball in the unlubricated channel is well below the bottom of the error bar for the lubricated channel. Therefore it appears the lubricating the channel does indeed change the time needed for the ball to roll down, and makes the time larger. The data allow one to reject the null hypothesis, and conclude that lubricating the channel has a significant effect on the time needed for the ball to roll.
6. To make this conclusion more precise, compare the data for lubricated and unlubricated channels with a t test (Equation 3.68). Using a two-sided test assuming equal variances, the probability of chance alone producing values as different as those appearing in Figure 2.1 is $p = 0.014$. This value of p is much less than the conventional value of $p = 0.05$ below which one can conventionally reject the null hypothesis. So the t test confirms that the experiment has arrived at a significant result.

No need to spend $100,000 on fancy equipment after all. The experiment has provided meaningful results despite the relatively large fluctuations in the individual measurements.

Fish in two lakes

In principle, one really would like to find the lengths of absolutely all the fish spines in Lakes Dugout and Ormond, take the averages of the two sets of lengths, and compare them. There is a finite number of fish in each lake, so this is a completely concrete question. The data give the spine lengths only of a small fraction of all the fish; they provide a *sample*. The mathematical procedure in Section 2.1.5 is just as appropriate here as it was for the rolling ball. In this case it provides a best guess of

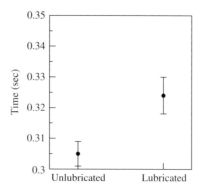

Figure 2.1 The experimental data in Table 2.2 are summarized by a plot of two averages and associated standard errors as computed in Equations (2.7) and (2.8).

the true mean length of fish spines based upon the sample – the sample average – plus an estimate of how much in error this best guess is likely to be – the standard error.

1. Finding the sample means of the fish spines by hand is tedious (although maybe not as tedious as collecting all the fish). A spreadsheet program can do the job quickly with the formulas =AVERAGE(B1:D35) and =AVERAGE(F1:I31) shown in Figure 2.2. In Ormond, the average fish spine length is $\bar{x}_{\text{Ormond}} = 4.92$ mm, while in Dugout it is $\bar{x}_{\text{Dugout}} = 4.67$ mm. This result is in accord with the working hypothesis, but it is not yet clear if the result is significant or just due to random variations in the samples.
2. Similarly, standard deviations can be computed in a spreadsheet through the cells =STDEV(B1:D35) and =STDEV(F1:I31).
   The results are

$$s_{\text{Ormond}} = 0.76 \text{ mm, and } s_{\text{Dugout}} = 0.57 \text{mm.}$$

This means that fish spines in the first lake differ characteristically by 0.76 mm from the average, and in the second lake by 0.57 mm from the average.
3. From the preceding cells, find the standard error through
   =B39/SQRT(COUNT(B1:D35)) and =F39/SQRT(COUNT(F1:I31))

$$\Delta x_{\text{Ormond}} = \frac{s_{\text{Ormond}}}{\sqrt{N}_{\text{Ormond}}} = 0.07 \text{ mm, and } \Delta x_{\text{Dugout}} = \frac{s_{\text{Dugout}}}{\sqrt{N}_{\text{Dugout}}} = 0.05 \text{ mm.}$$

This means that sampling errors are likely to cause the sample average in Ormond to differ by around 0.07 mm from the true average, and by around 0.05 mm in Dugout. The results so far can be summarized by

$$x_{\text{Ormond}} = 4.92 \pm 0.07 \text{ mm}; \ x_{\text{Dugout}} = 4.67 \pm 0.05 \text{ mm.} \qquad (2.9)$$

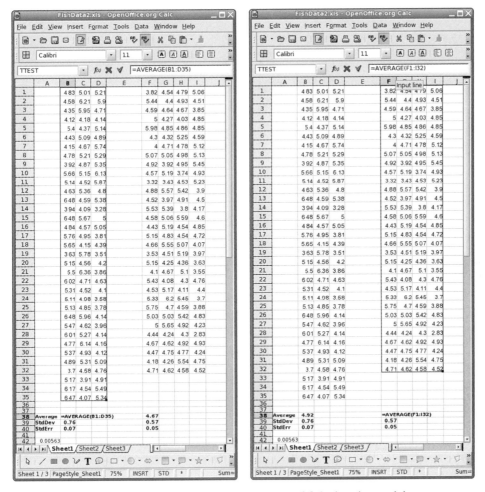

Figure 2.2 Computing the sample averages of fish sizes in two lakes.

4. A graph of the results so far appears in Figure 2.3.
5. The bottom of the error bar for fish spines in Ormond is well above the top of the error bar for fish in Dugout. Therefore it appears that fish spines in Ormond are really larger on average than those in Dugout. Despite sampling errors, the data allow one to reject the null hypothesis, and are consistent with the idea that the presence of predators should lead to larger fish spines.
6. To verify the conclusions reached by inspecting the graph, run a two-tailed t test, assuming unequal variances. The result is $p = 0.0056$, which is much less than the conventional threshhold of $p < 0.05$ for concluding that results are significant.

Sampling error is not a problem after all, and one can feel quite confident in saying that the fish in Ormond are larger than those in Dugout.

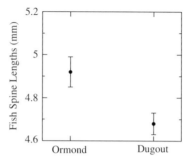

Figure 2.3 The experimental data from Table 2.3 are summarized by a plot of two averages and associated standard errors as computed in Equation (2.9).

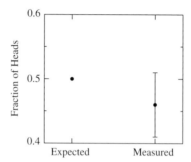

Figure 2.4 The experimental data from Table 2.4 are summarized by comparing the expected value of 0.5 for fraction of heads with the measured average value of 0.46.

Flipping a coin

The property of the coin one wants to investigate is its "fairness." The definition of fairness is that if flipped infinitely often, the coin comes up heads exactly half the time. More precisely, it is that in the limit of infinite flips, the fraction of heads converges to 0.5. In this case, the issue to settle is whether one can tell if a coin is fair after 100 flips, and whether having 46 heads instead of 50 indicates that it is not.

1. In Table 2.4 there are 46 1's and 54 0's, so the average is 0.46.
2. The sample standard deviation is $s = 0.501$. It should come out to be very close to 0.5 because each number in the table is either 1 or 0, the average is very close to 0.5, and therefore each individual value deviates by approximately a magnitude of 0.5 from the average. This is the definition of the standard deviation.
3. The standard error is $\Delta x = s/\sqrt{100} = 0.05$.
4. A graph of the result appears in Figure 2.4.
5. The error bar overlaps the expected value of 0.5. Therefore one has no grounds to reject the null hypothesis, and so far as one can tell the coin is fair. Finding 54 tails and 46 heads in 100 flips is quite consistent with the random nature of flipping a fair coin.

Table 2.5 Two types of possible error in testing a hypothesis.

	Null hypothesis is really **true**	Null hypothesis is really **false**
Researcher **accepts** null hypothesis, rejects alternative	Correct	Type II error (false negative)
Researcher **rejects** null hypothesis, accepts alternative	Type I error (false positive)	Correct

6. A statistical test that compares a set of measurements with an expected value is the Z test. Running this test gives a value of $p = 0.42$, which is larger than 0.05, and which means that chance alone could easily be expected to produce an average value that differs this much from the expected value of 0.5.

The bottom line is that so far as one can tell from 100 flips, the coin is fair. The coin might still be somewhat unfair and this experiment would not show it. If in the limit of an infinite number of flips, the coin comes up heads only 49% of the time, it is possible to detect, but many more flips are needed. Section 2.3 and an example on page 91 will describe how to design a new experiment in cases like this.

2.1.7 Types of errors

There are two different types of errors that can arise when testing a hypothesis, as shown in Table 2.5. In a *Type I error*, the researcher rejects the null hypothesis, even though it is correct. That is, the researcher believes there is an effect when there is none. This sort of error is also called a *false positive.* An example is Linus Pauling and the prevention of colds by vitamin C (Pauling, 1971; 1986). Pauling won two Nobel Prizes, one in Chemistry and the other in Peace for nuclear disarmament, so his stature was enormous. He spent much of the latter part of his life pressing the idea that vitamin C had great health benefits, among which was the ability to prevent common colds and shorten their duration. Decades of increasingly accurate experiments simply could not find a connection between taking large doses of vitamin C and cold prevention (Douglas and Hemilä, 2005). The current scientific belief is that there is no connection. Pauling's influence is still felt because millions of people take vitamin C in response to colds anyway.

In a *Type II error,* researchers accept the null hypothesis when it is false. That is, there really is an effect, and the researcher denies it. This sort of error is also called a *false negative.* There were many arguments about this type of error when smoking was first suspected of causing lung cancer and heart disease. Eventually, the connection between smoking and heart disease became too clear to deny. Even so, the arguments continued. Sir Ronald Fisher, then the world's greatest living statistician and a smoker, offered the "constitution hypothesis;" that people might genetically be disposed to develop these diseases and to smoke. The matter was finally put to rest by studies, such as by Kaprio and Koskenvuo (1989) and Calori *et al.* (1996), of identical twins, one of whom smoked and the other did not. The twins who smoked had much higher rates of heart disease and lung cancer.

2.2 Measuring values

Whether as a component of a more elaborate experiment, or as a goal in itself, measuring an experimental value is one of the most common scientific tasks. Every attempt to measure a value must deal with random and systematic errors. Random errors are those that vary from trial to trial and whose effect can be eliminated by sufficient averaging. Systematic errors produce the same effect in every measurement and can only be treated through improved experimental design or analysis. Consider each type in more detail.

2.2.1 Random error

Suppose one is trying to measure something as apparently straightforward as the mass M of a single penny. The last digit on the scale flips about randomly because of fluctuations in electrical current inside the scale, or because of small air currents that hit the plate where the penny sits. Factors such as this are called *noise* and contribute to *instrumental or random error*. The important point is that this sort of error can be eliminated completely by taking a sufficient number of independent measurements. Every particular measurement is influenced by noise, but the contribution of noise to the average of the measurements vanishes for sufficiently many measurements. The way averaging works has been summarized in Section 2.1.5.

To be more specific, consider taking 10 measurements, which in grams are listed in Table 2.6.

Following the procedure of Section 2.1.5 and using Equations (3.1) and (3.7), the sample average is 2.529, the sample standard deviation is $s = 3.11 \times 10^{-3}$, and the standard error, or uncertainty in the mean, is $s/\sqrt{10} = 9.9 \times 10^{-4}$. Rounding the standard error up to 0.001, one says that the mass of the coin is

$$M = 2.529 \pm 0.001 \text{ grams.}$$

Table 2.6 Data for weight of a penny.

2.532	2.525	2.526	2.531	2.534
2.527	2.525	2.528	2.529	2.531

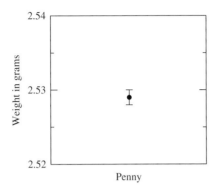

Figure 2.5 Mass of a penny indicated through a dot for the sample mean and error bars for the standard error of the mean.

There is no sense in retaining lots of decimal places to describe the standard error.

In drawing a graph, it is traditional in physics and chemistry to indicate 2.529 ± 0.001 as shown in Figure 2.5. Not every community has exactly the same convention regarding error bars. Sometimes error bars can be used to indicate two standard errors about the mean. If there is any doubt, the caption of the figure should specify the convention that is being used.

2.2.2 Systematic error

Some sorts of error cannot be cured by repeated measurement. Suppose someone gives you a plastic ruler to measure the width of a penny. What they have not told you is that they put it in an oven, and it has shrunk by 10%. Every measurement you make gives a larger value than it should, and the errors do not cancel out with averaging. Every measurement you make is off, and using the ruler over and over again does not help in the least.

In the case of random error, there are definite rules for reducing its consequences. One makes measurements repeatedly and averages them together. For systematic error, there is no procedure guaranteed to solve the problem. However, there is a process that will catch many forms of systematic error, the process of *calibration*.

Equipment calibration worksheet

Your time and our equipment are valuable. Yet many of the experiments performed using probes and other laboratory equipment are unsuccessful because the equipment is improperly prepared and used. Find the manual. Read the manual. Ask questions of the instructors and TAs before proceeding to experimentation. *Complete the following steps before first using the instrument and if appropriate every time before using the instrument:*

- What does the instrument measure? Be specific, including the units of measurement.

- How does the instrument take measurements; i.e., how does the probe collect the data?

- What is the maximum value the instrument can measure?

- What is the minimum value the instrument can measure?

- What is the instrumental limit of error?

- Note below any steps that must be performed to prepare the instrument for use.

- Note below any conditions that must be maintained while the instrument is in use.

- Briefly describe the calibration you will perform. This means that you use the instrument in a case where the answer you should obtain is known.

- Note the value you obtained in the calibration:

- If you are taking your equipment out of the classroom, and if the measurements you will take differ in any fashion from the calibration measurement you have performed, you must carry out a trial run before leaving the classroom with equipment.

The most common form of systematic error comes from measuring instruments that consistently deliver incorrect values. The essential idea of calibration is to check the measuring instrument as carefully as possible. The Equipment Calibration Worksheet on page 32 contains a range of questions one should be able to answer about every measuring instrument. One should know the maximum and minimum values it is able to measure. One should know the smallest values it is able to tell apart (the instrumental limit of error). One should know about any limitations on how the instrument can be used. Some probes are meant to be used in a vertical position, and give nonsense when turned sideways. Some stop working outside a certain range of temperatures. Almost any instrument more complex than a ruler should come with a user's manual that describes precautions such as these. So you should read the user's manual for every scientific instrument before you use it.

Even if all the directions in the user's manual are followed faithfully, the instrument may still be broken. Springs in scale balances can bend, membranes in chemical probes can clog or tear. The only precaution one can take is to use the instrument to measure some quantity whose value is quite certain. This known quantity is called a *standard*. For example, pH probes come with solutions of known pH that need to be used in a calibration process before every use. In a *two-point calibration*, a measuring instrument is set to reproduce a standard at two separate points, one toward the high side of its range, one towards the low side of its range. Even measuring scales can be checked against pre-measured standard weights. It is better to have a rough calibration than none at all. So if an oxygen sensor used in the ordinary atmosphere reports measuring 5% oxygen, the sensor is certainly defective, since if there were so little oxygen, everyone nearby would be dead. One can check that an oxygen sensor gives a value of around 22% oxygen in the atmosphere even if there is no special bottle available with precisely 21.93% oxygen to check exactly.

In catching systematic errors, it is helpful to draw on a range of reliable known values – the density of water, the acceleration of gravity, typical values for the chemical composition of the atmosphere.

When scientists perform experiments containing systematic errors and publish them, the process of finding the errors can be extremely slow. Sometimes the errors may be caught by other scientists with expert knowledge during the review process. Sometimes they may be revealed when other scientists repeat the measurements using different methods. Based on the past history of finding errors, there must be many incorrect results currently believed to be true (one hopes there are not too many that are very important).

Here is a famous example of a systematic error that was not caught for many years (Figure 2.6). Robert Millikan was one of the most famous American

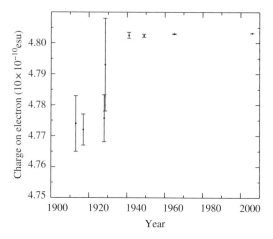

Figure 2.6 Charge on electron versus time. Data prior to 2008 taken from a sequence of review articles in *Reviews of Modern Physics* on fundamental physical constants by Birge, and Dumond and Cohen, such as Cohen and DuMond (1965). Current data from NIST (2008).

physicists at the beginning of the twentieth century. The work with which he made his name and for which he won a Nobel Prize was the measurement of the charge on the electron, carried out between 1910 and 1920 at the University of Chicago and Cal Tech. Previous experiments detecting the electron found the ratio of its charge to its mass. Millikan developed a new technique where oil droplets acquired charges of single electrons and were placed in electric fields. The speed at which the oil droplets moved while gravity pulled them down and electric fields pulled them up made it possible to deduce the charge of the electron. However, in order to find this fundamental constant, it was necessary to know also the precise amount of drag that air exerted on the oil droplets when they started moving. This required very precise measurements of the viscosity of air. In fact, the results were sensitive enough to the viscosity of air that it was necessary to take into account temperature corrections, since the viscosity of air had been measured at one temperature, and Millikan's experiments were carried out at a temperature a few degrees different. Unfortunately, published numerical values for how much the viscosity of air changed with temperature were wrong. The net result was that the charge on the electron came out 1% too low. Meanwhile, Millikan's error estimates said that the results should be accurate within 0.2%. The consequence of systematic errors is that error estimates themselves are in error. Not that Millikan's experiment or results were bad. They are still regarded as some of the most ingenious and important experiments in the history of science. Obtaining the charge of the electron even with 1% accuracy was a wonderful advance. However, the error in estimating the error was a blemish. The problem was not corrected until 1928, when a Swedish graduate student, Wadlund (1928), found a completely new method to measure the charge of the electron. The new method combined measurements of the rate that electrical currents deposited silver on an electrode with measurements of X-rays bouncing off calcite crystals. Wadlund at first put huge error bars on his measurements, maybe in terror at the thought of disagreeing with Millikan. Within a few years, numerous researchers duplicated measurements using the new method, found that they were more accurate than Millikan's, and that the two methods were not in complete accord. It took several years more before the problem was traced to the temperature corrections for the viscosity of air. Since 1928 there has been steady improvement in the accuracy with which the charge on the electron has been measured. Each improvement has fallen within the error bars of the previous measurements. The current value is $4.803205 \times 10^{-10} \pm 4 \times 10^{-17}$ esu.

2.2.3 Precision and accuracy

The ideas of *precision* and *accuracy* give an additional way of thinking about systematic error. A numerical result is very precise when it is specified to many digits, or when the error bars surrounding it are very small. Put the numbers for the weight of a penny in Table 2.6 into a calculator, calculate the average, and the calculator will return the number 2.5287999999999995. This is a very precise result because it specifies the answer to one part in 10^{16}. But this precision is quite misleading, because the weight of the penny is not actually known with that much accuracy. The accuracy with which the result is actually is known is ± 0.001.

In any experiment where there are systematic errors, there will also be a difference between precision and accuracy. By repeating an experiment a very large number of times, the standard error can be made as small as desired. In the case of Figure 2.6, Millikan repeated his experiment often enough that he thought the results were precise to within $\pm 0.005 \times 10^{-10}$ esu. However, because of systematic error, they were only accurate within 0.03×10^{-10} esu.

In scientific writing, you should record just enough digits so that the precision with which you specify a result and the accuracy with which you know it are about the same. Suppose you measure a stack of 29 sheets of paper and find that it is 3.5 mm thick. When you tell this result to someone, they will probably conclude that you were able to determine the thickness to within around half a millimeter, which is correct if you have done the measurement with an ordinary ruler. It is better to indicate your uncertainty explicitly by reporting 3.5±0.5 mm, since you might have been able to use a caliper that could provide accuracy of 0.1 mm. Next you may want to know the thickness of a single sheet of paper, and dividing 3.5 by 29, your calculator presents you the result 0.12068965517241 mm. You should not report this full number, because it implies you know the answer to 13 places of precision. In fact, the certainty with which you know the thickness of the single piece of paper is 0.5 mm / 29 ≈ 0.02 mm, and you should report the thickness of a piece of paper as 0.12 ± 0.02 mm. In short, in current scientific practice it is wrong to record all the digits your calculator or computer gives you in writing up a research project, because the calculator is able to provide precision that is much greater than the accuracy of your measurement.

2.2.4 Propagation of error

Frequently, some quantity you wish to know can only be obtained by combining together a number of different measurements. If each of these measurements has some uncertainty associated with it, either because of limitations in measuring devices, or because of uncertainty in a sample average (standard error of the mean), you will need to find how the uncertainty of your final result depends upon uncertainties in the numbers that go into it. This problem is solved through *propagation of error*.

As an example, suppose that instead of measuring the weight of a penny, you set out to measure its density, which is mass divided by volume. Finding the volume of a penny accurately is not so easy because it is a small object with a complex shape, including the engravings of Lincoln and the Lincoln Memorial, but one can get an estimate using an ordinary ruler. The diameter of the penny is

$$d = 18.5 \pm 0.25 \text{ mm.} \tag{2.10}$$

The error of 0.025 cm comes not from repeated measurement and averaging, but from an estimate of the maximum accuracy with which the length of the penny can be read off by eye. The smallest division on a normal metric ruler is 1 mm; this is the *least count*. One can distinguish fractions of this distance by eye, depending on how good ones eyes happen to be. The smallest values one can safely distinguish with an instrument is called the *instrumental limit of error*, which in this case is taken to be 0.25 mm. The thickness of the penny is

$$t = 1.25 \pm 0.25 \text{ mm.} \tag{2.11}$$

Once again, the instrumental limit of error is used to indicate the uncertainty in the measurement. In addition, the width of the penny varies from spot to spot because of engravings, and there is a lip at the edge where the width is measured that makes the penny seem a bit wider than it must be on average.

Neglecting the engravings of Lincoln and his Memorial to find the volume of the penny, use the formula for the volume of a cylinder

$$V_{\text{cylinder}} = \pi \left(\frac{d}{2}\right)^2 t. \tag{2.12}$$

This computation would be easy, except that both the diameter and thickness of the penny have associated uncertainties. These lead to some uncertainty in the volume of the penny, which will finally contribute to uncertainty in the value of its density.

The central result, which is surprising, is that when two different quantities with independent uncertainties are added together, the uncertainty of the sum is **not** the sum of uncertainties. Instead, if one wants to add two sample means $\bar{x} \pm \Delta x$ and $\bar{y} \pm \Delta y$ then the uncertainty in the sum $\bar{x} + \bar{y}$ is $\sqrt{\Delta x^2 + \Delta y^2}$ (see Equation 3.62 on page 80). This result can be written as

$$\bar{x} \pm \Delta x \; + \; \bar{y} \pm \Delta y \; = \; \bar{x} + \bar{y} \; \pm \; \sqrt{\Delta x^2 + \Delta y^2}. \tag{2.13}$$

In the particular case where the uncertainties Δx and Δy are of the same size, Equation (2.13) says that the uncertainty in their sum is $\sqrt{2}$ times as big as the individual uncertainties, not twice as large.

Through repeated use, this one formula can be employed to show how uncertainties propagate in general. As a first example, consider adding three quantities, $\bar{x}$, $\bar{y}$, and $\bar{z}$. One has

$$\bar{x} \pm \Delta x \; + \; \bar{y} \pm \Delta y \; + \; \bar{z} \pm \Delta z = \left(\bar{x} + \bar{y} \pm \sqrt{\Delta x^2 + \Delta y^2} \right) + \bar{z} \pm \Delta z$$
$$= \bar{x} + \bar{y} + \bar{z} \pm \sqrt{\Delta x^2 + \Delta y^2 + \Delta z^2}. \tag{2.14}$$

As a second example, consider multiplying two uncertain quantities together.

$$(\bar{x} \pm \Delta x) \; (\bar{y} \pm \Delta y) = \bar{x}\bar{y} \left(1 \pm \frac{\Delta x}{\bar{x}} \right) \left(1 \pm \frac{\Delta y}{\bar{y}} \right)$$
$$= \bar{x}\bar{y} \left(1 \pm \frac{\Delta x}{\bar{x}} \pm \frac{\Delta y}{\bar{y}} + \text{small quantities} \right). \tag{2.15}$$

Assuming that the uncertainties Δx and Δy are small compared to $\bar{x}$ and $\bar{y}$ (this is very much the most common case) the product of the uncertainties Δx and Δy should be so small as to be negligible. Dropping a last term that would otherwise be present in Equation (2.15), one can use Equation (2.13) to conclude that

$$(\bar{x} \pm \Delta x) \; (\bar{y} \pm \Delta y) = \bar{x}\bar{y} \left(1 \pm \sqrt{\left(\frac{\Delta x}{\bar{x}}\right)^2 + \left(\frac{\Delta y}{\bar{y}}\right)^2} \right) \tag{2.16}$$

$$= \bar{x}\bar{y} \pm \bar{x}\bar{y} \sqrt{\left(\frac{\Delta x}{\bar{x}}\right)^2 + \left(\frac{\Delta y}{\bar{y}}\right)^2}. \tag{2.17}$$

More examples of error propagation appear in Table 2.7. They can be used one after the other when formulas involve combinations of addition, powers, and products. It

Table 2.7 Formulas for propagation of uncertainty.

Sum two uncertain quantities	$\bar{x} \pm \Delta x + \bar{y} \pm \Delta y$	$\bar{x} + \bar{y} \pm \sqrt{\Delta x^2 + \Delta y^2}$
Sum many uncertain quantities	$\sum_i (\bar{x}_i \pm \Delta x_i)$	$\left(\sum_i \bar{x}_i\right) \pm \sqrt{\sum_i \Delta x_i^2}$
Multiplication by constant	$c(\bar{x} \pm \Delta x)$	$c\bar{x} \pm c\Delta x$
Power	$(\bar{x} \pm \Delta x)^n$	$\bar{x}^n \pm |n|\Delta x$
Multiplication of two uncertain quantities	$(\bar{x} \pm \Delta x)(\bar{y} \pm \Delta y)$	$\bar{x}\bar{y} \pm \bar{x}\bar{y}\sqrt{\left(\frac{\Delta x}{\bar{x}}\right)^2 + \left(\frac{\Delta y}{\bar{y}}\right)^2}$
Division of two uncertain quantities	$(\bar{x} \pm \Delta x)/(\bar{y} \pm \Delta y)$	$\bar{x}/\bar{y} \pm \bar{x}/\bar{y}\sqrt{\left(\frac{\Delta x}{\bar{x}}\right)^2 + \left(\frac{\Delta y}{\bar{y}}\right)^2}$

may appear there is a contradiction between the formulas. For example, according to the formula for multiplying by a constant,

$$2(\bar{x} \pm \Delta x) = \bar{x} \pm \Delta x + \bar{x} \pm \Delta x = 2\bar{x} \pm 2\Delta x,$$

and this statement seems to contradict the claim that adding two errors produces an uncertainty $\sqrt{2}$ times larger than the original. The reason is that the errors being added must be completely uncorrelated with one another for the result to be $\sqrt{2}$ times as big. When the error in one quantity is positive, half the time the error in the other needs to be negative. It is this possibility of cancellation that leads to a factor of $\sqrt{2}$. But when one adds the uncertainty in $\bar{x}$ to the uncertainty in $\bar{x}$, there is no possibility of cancellation. The errors are always exactly equal to each other. Because they are correlated, they add up through normal rules of addition rather than according to the peculiar rule in Equation (2.13).

Returning to the example of finding the density of a coin, one first needs to compute the volume of the coin. The diameter d is squared in Equation (2.12), and using the expression for a power in Table 2.7 gives

$$d^2 = (18.5 \pm 0.25 \text{ mm})^2 = 342 \pm 0.5 \text{ mm}^2. \tag{2.18}$$

Next, multiply by the thickness t, using the rule for multiplication,

$$d^2 t = (342 \pm 0.5)\text{mm}^2 (1.25 \pm 0.25) \text{ mm} \tag{2.19}$$

$$= 428 \left(1 \pm \sqrt{\left(\frac{0.5}{342}\right)^2 + \left(\frac{0.25}{1.25}\right)^2}\right) = 428 \pm 85 \text{ mm}^3. \tag{2.20}$$

Using the rule for multiplication by a constant gives

$$V_{\text{penny}} = \frac{\pi}{4}d^2 t = 336 \pm 67 \text{ mm}^3 = 0.336 \pm 0.067 \text{ cm}^3. \tag{2.21}$$

The uncertainty in the volume is around 20% and mainly comes from the uncertainty in the thickness of the penny. Finally, using the expression for division, the density ρ of the penny is

$$\rho = \frac{M}{V} = \frac{2.529 \pm 0.001 \, \text{grams}}{0.336 \pm 0.067 \, \text{cm}^3} = 7.53 \left(1 \pm \sqrt{\left(\frac{0.001}{2.529}\right)^2 + \left(\frac{0.067}{0.336}\right)^2} \right) \frac{\text{grams}}{\text{cm}^3}$$

$$= 7.53(1 \pm 0.2) \frac{\text{grams}}{\text{cm}^3} = 7.5 \pm 1.5 \frac{\text{grams}}{\text{cm}^3}.$$

This final result for the uncertainty would have come out the same if the only error employed was the uncertainty in the thickness of the penny. If there is a single quantity in a formula that is much less certain than the others, its uncertainty dominates the final result.

2.3 Improving experiments

2.3.1 Improving hypothesis-driven research

Clever Hans could add numbers together and do simple calculations involving dates. This would not have been remarkable, except Hans was a horse and he answered questions by tapping on the ground. Through the 1890s his owner Wilhelm Von Osten paraded him throughout Germany. An 11-member commission established to investigate Hans' remarkable intelligence could find no fraud or tricks and in 1904 concluded that Wilhelm Von Osten had successfully used with his horse the methods he used to teach arithmetic to school children. So addition and subtraction by horses might be well accepted today, except for the further investigations of psychologist Oskar Pfungst. Pfungst found that Hans could answer questions correctly only when the questioner knew the answer, and when Hans could see the questioner. The horse stopped knowing arithmetic when he was made to wear blinders, and the person asking him questions stood to the side. Anyone asking the horse a question would involuntarily jerk their heads slightly upwards when the number of taps reached the correct value. The horse noticed the signal and stopped tapping. Before Pfungst used blinders, person after person tested the horse without realizing that their expectation of the right answer biased the result (Pfungst, 1965).

Bias is a general problem of hypothesis-driven research. Bias refers to actions taken consciously or unconsciously by the experimenter that damage the validity of the experiment, and is a form of systematic error. One form it can take arises when the amount of data to be gathered is not settled in advance. The researcher goes along, checking whether the answer is an agreement with an expected value, and when it gets close enough, he stops taking data. To protect against this sort of bias, you should settle on a certain amount of data collection, gather the data you have planned, and then do the analysis.

In research involving people, particularly medical research, there is a great risk that expectations about treatment will bias the results. To avoid these effects, the experiments are designed to be *single-blind* or *double-blind*. Experiments of this sort often involve a new treatment, such as a new drug or a new form of surgery. Patients receiving the new drug are the *treatment group* and those getting the old drug, or no drug are the *control group*. The question is whether the new treatment is superior to an older established one. The null hypothesis is that it is not. In a single-blind experiment, the subjects receiving treatment are not told whether they are receiving the old or new treatment. However, the people administering the experiment know who is getting what, and there is still a risk of bias because of unconscious signals that people delivering treatment may give to the patients. In a double-blind experiment, neither the subjects nor the people administering treatment know who is receiving the old or new medicine, and they are not allowed to find out until the data have been gathered.

Sometimes an experiment needs to examine whether taking some drug is better or worse than taking nothing at all. To keep subjects in the dark about whether they are in treatment or control groups, some of the patients can be given a *placebo* which is a pill that looks like the real medicine but contains completely harmless ingredients. One of the major attempts to study the effectiveness of vitamin C in preventing colds was damaged because many of the subjects began sucking on their placebo pills and realized they were sugar and not vitamin C (Karlowski *et al.*, 1975).

Bias can also threaten laboratory experiments, particularly when the quantities being measured are difficult to detect and some judgment is needed to determine whether they have been seen or not. Blinding procedures have become customary in large physics experiments, where the success or failure of an enormous international effort hinges upon finding some very rare events (Klein and Roodman, 2005). One way to remove bias in these cases is to give researchers a small random subset of the data, have them design and carry out the analysis on this subset, and then insist that analysis of the complete data set be carried out in exactly the same way.

2.3.2 Improving the measurement of values

Laboratory experiments

Since the main difficulties in measuring experimental values come from random and systematic error, improving measurements means reducing these errors. In the case of random errors there are two basic ways. The first is to redesign the experiment so as to reduce the amount of noise. For the example of the ball running down lubricated and unlubricated ramps, this could mean sanding down the ramp (see Galileo's discussion of a ball rolling experiment in Appendix B), or building a machine to release the ball at the top of the ramp rather than releasing it by hand. The second way is to take more data, as the effects of randomness eventually disappear when one takes a large enough sample of data. But please be aware that the effects of randomness disappear quite slowly, and one may have to take a **lot** of data.

Just how much data one has to take comes from the expression for the standard error of the mean which says that

$$\Delta x = \frac{s}{\sqrt{N}}. \tag{2.22}$$

As an experimenter takes more and more data, the sample standard deviation s settles down to a fairly constant value, representing the typical amount by which individual measurements differ from the mean. In contrast, the standard error Δx decreases as the sample size N increases. Mathematically, this happens because of the factor of $\sqrt{N}$ in the denominator of (2.22). Conceptually, it happens because the more measurements one takes, the less uncertainty due to random errors is left. Unfortunately, the standard error does not decrease as fast as one would like. To see the sample sizes this equation recommends, solve (2.22) for N to find

$$N = \left(\frac{s}{\Delta x}\right)^2. \tag{2.23}$$

What Equation (2.23) means is that if one carries out an experiment and decides to take more data to make the error bars smaller, the number of data points required goes as 1 over the square of the desired reduction in size of the error bars. Return to the example of flipping a coin on page 28 where the standard error came out to be 0.05. Suppose one decides to pin down the fairness of the coin 10 times better, and wants the standard error to be 0.005 instead. The sample standard deviation was almost exactly 0.5, so inserting $s = 0.5$ and $\Delta x = 0.005$ into Equation (2.23) gives $N = 10{,}000$. Decreasing uncertainty by a factor of 10 requires increasing the number of measurements by a factor of $10^2 = 100$.

To reduce systematic error, all one can do is to calibrate instruments carefully, or design the experiment using new and more accurate instruments. For example, in finding the density of a penny, the accuracy could be increased by using a caliper in good condition to measure the thickness of the penny rather than an ordinary ruler.

2.3.3 Sampling and surveys

In experiments such as the comparison of fish spine lengths in Section 2.1.6, measuring a subset or sample will only represent well all the fish in the lake if the sample is *random* and *representative*. The fish should come equally from all parts of the lake, and not just from one spot near one shore.

The need to sample populations in a random and representative way is particularly important in the case of *surveys*. A survey might ask people for their intention to vote for a particular candidate, their opinion on some public issue; a test might assess their knowledge in math or science. Almost always the intention of a survey is to learn by sampling a small number of people about a much larger group. For example, political polls typically ask between a few hundred and a few thousand people which candidate they support. No one particularly cares who that tiny group supports; the point is that they are supposed to be representative of a much larger group such as the full voting public in the U.S. Conclusions based on the poll will only be legitimate if the sample is really drawn randomly from the group one wants to learn about. Every poll is reported with a margin of error, which is usually defined to be roughly twice the standard error of the mean. If a poll says that 48% of the public plans to choose Harold Belleview for President with a margin of error of ± 4 percentage points, this means that the result of the poll was that the sample mean $\bar{x} = 0.48$ and the true answer has a high probability to lie in the range $48 \pm 4\%$.

The sorts of opinion polls that are so familiar on television and in newspapers today gained their authority by going head to head with less scientific methods and winning. The turning point for scientific polling came in the 1936 Presidential election between Roosevelt and Landon. George Gallup Jr., son of the founder of the Gallup polling organization, describes it this way (Gallup, 2000):

In 1936, in the election year, my dad had the temerity to challenge the Literary Digest, which had developed an incredible record since the turn of the century in predicting elections. My dad had the temerity to predict how far off the mark they would be. You will have to check these figures, but he said that they'd be 19 points off, and they were 18 points [off]. So that was more accurate, that prediction, than our own at that time, but he and Crossley and Roper went on to predict that FDR would win.

Now, the reason that the Literary Digest was way off the mark is that they had developed their sample from lists of car owners and telephone owners. And these tended to be more upscale than the average person. And up to that time, political lines didn't follow economic lines particularly. But in that race in 1936, political lines followed economic lines very sharply. So the Literary Digest in going to upscale people got a much more Republican figure. My dad attempted to get people of all political backgrounds, upscale and down scale, and so they were on the mark. The Literary Digest was way off the mark. The Gallup operation and, as I mentioned, Crossley and Roper were calling FDR the victor.

The Literary Digest also assumed that the more people you interviewed, the closer you're going to get to the truth, the actual, and of course that's not so. As a matter of fact, they went to about a third of all households in the United States. And the assumption was, the more people you interview, of course, you're going closer to the truth. But, of course, the key is, whom are you interviewing, the characteristics of those people. . . .

To find people we were using what we call the quota sample in those days. My dad had, after a great deal of experimentation – in fact he looked at the effect of fifty-four different questions in terms of demographics – determined that there were five key determinants of opinion, and they were age and sex, income, region, and so forth. And then he would send his interviewers out to get a quota of people from each of those groups.

It takes the resources of a national organization to obtain a representative random sample of U.S. citizens. If you want to carry out such a survey as part of a project, this will probably not be possible. By being thoughtful, you can increase the chance that a survey you perform is meaningful. For example, suppose you want to carry out a survey to find the percent of U.S. undergraduates who accept the theory of evolution. This would require asking questions of undergraduates across the entire U.S. Hard! It is much more reasonable to attempt to find the opinions of under-graduates at your own university. Ideally, you would obtain a list of all registered undergraduates, choose a completely random sample, contact them and administer the survey. There would be the potential of some bias in who responded depending upon the form of contact. One way to obtain random lists of student names and con-tact information is to type two-letter sequences into a university directory search engine. This process also allows you to select students by college or major. The problem with this method is that the response rate to a survey of this sort carried out by a student is likely to be very low. You may have better luck going to some location such as the student union or a library and asking for volunteers. If you do this, then you must make every effort to understand how the physical location biases the sample. If you go to the life sciences library and poll students on their thoughts concerning evolution, you will probably get very different results than if you went to the math library or the music library. Perhaps the best you will be able to do will be to understand and describe the weaknesses in the procedure you have followed.

2.4 Safety and ethics

One of the reasons that scientists have a poor image in many movies and television shows is that they are portrayed as pursuing scientific questions without any regard to human values. There have been scientists who acted in this way, and some scien-tific creations such as the nuclear bomb will always be controversial. However any scientific experiment that does not pay complete attention to the health and safety of all participants is unethical, should not be carried out, will be banned from publication, and condemned by all scientific organizations and other scientists.

2.4.1 Laboratory safety

If you are working in a laboratory, you must respect your own safety, and the safety of those working around you. You must come to lab appropriately dressed and be aware of emergency equipment available to you in case of accidents. Appendix C describes safety requirements in more detail. Some of the greatest risks to safety include fire, spilled chemicals, broken glassware, and power tools used without needed caution. You can also injure people around you by jostling them, or even by making loud unexpected noises.

2.4.2 Animal safety

If your experiment involves animals, you are obliged to treat them humanely when possible and avoid pain and death except as justified by benefits to humans. The degree to which these principles must be followed depends upon the type of animal. The U.S. Federal government has established standards for the treatment of animals in the the Animal Welfare Act, and in a university setting experiments involving animals must be reviewed by the Institutional Animal Care and Use Committee. The review becomes particularly careful and strict if pain, sickness, or mortality are involved. An outline of your responsibility to animals according to their type is:

Invertebrates

There are minimal restrictions on how you can treat invertebrate animals such as insects. You do not need to have such experiments reviewed by the Institutional Animal Care and Use Committee. However you should not inflict needless pain and suffering on any animal in the name of research.

Vertebrates

You must treat vertebrates in a humane fashion, and you cannot kill them or inflict pain without careful review and justification. If you do conduct an experiment that causes pain, sickness, or death to vertebrate animals, the suffering of the animals must be justified by benefits to humans. In the case of primates, the mental health of the animals must be considered as well. Not all philosophers agree that this trade-off is ethical (Singer, 1975), but it is allowed currently under U.S. law.

Any experiments involving vertebrates in a university setting must be reviewed by the Institutional Animal Care and Use Committee. It is unlikely that you will have time for review and approval during this class. Therefore you should restrict yourself to observational studies of animals in a natural setting, or experiments involving pets that involve activities you would perform normally while taking care of them. No experiments with vertebrate animals that could cause pain or threaten their health will be allowed.

2.4.3 Human subjects

The strictest protections of all govern the use of human subjects in research. Current U.S. laws respond to abuses carried out in the name of science in the recent past, particularly in Germany in the 1930s and 1940s, and in the United States in the 1950s. The National Research Act of 1974 established a commission given the task of articulating the primary principles governing use of humans in research, and the primary practical consequences of those principles. The resulting Belmont Report can be found at Belmont (1974). Three ethical principles singled out by the report are that:

1. All persons must be respected.
2. Researchers must attempt to increase the well-being of all they study, and at least avoid doing them harm.
3. Benefits and costs of research should be distributed in a just fashion.

The primary consequences drawn from these principles are:

1. All participants in a scientific research project must give informed consent. They must have adequate information on the experiment in which they participate, they must understand the information they are given, and they must participate willingly.
2. Every research project must be subject to an assessment of costs and benefits. Brutal treatment of humans is never justified in the name of research, and all risks to participants must always be minimized.
3. Research subjects must be selected in a just manner.

In a university setting, these requirements are enforced by the Institutional Review Board, a committee established to review every research project using human subjects. The review is careful and enforcement of ethical principles is strict. One of the principles is that if in the midst of an experiment there are conclusive data showing that the experimental procedure is harming the subjects, the experiment must be terminated immediately. Thus, in the fall of 2004, the drug company Merck immediately withdrew the arthritis and pain medicine VIOXX from the market, although its sales the previous year had exceeded $2 billion. The reason was that trials were underway to determine if VIOXX prevented colon polyps. Instead, results showed that the risk of suffering from heart attacks, strokes, and blood clots increased to 15 cases in three years per 1000 patients versus 7.5 cases in three years per 1000 patients for a control group taking a placebo.

Procedure to ensure voluntary participation and anonymity

To satisfy voluntary participation and anonymity simultaneously, here is a practical procedure. Create permissions forms such as those below. Every permission form must be individually numbered, and must match an individually numbered survey instrument. The surveys or tests and the permission forms must be on different pieces of paper. Each participant should sign one copy of the permissions form and give it to the researcher and keep a copy so as to know how to remove themselves from the study later if desired. The name of the participant must not appear on the survey instrument, and the researcher must pledge not to compare the names on the permissions forms with the responses on the survey at any time. The only exception occurs if the participant should contact the researcher after the survey has been performed and ask to be removed. In this case, the researcher finds the survey number from the permissions form, uses it to locate the survey, and destroys both the survey and permissions form. If analysis has been completed, the data must be removed from the analysis as if the participant had never agreed to begin with. Surveys and permission forms should be kept separately. Do not look at the permission forms when you tabulate and analyze the responses. Give both response forms and permission forms to your instructor once the analysis is complete.

Consent to Serve as a Subject in Research   Survey # 23
I am at least 18 years of age and consent to serve as a subject in this survey.

- I understand the purpose of this survey and have had opportunities to ask questions.
- My survey responses will be kept confidential and responses will not be publicly disseminated outside of a class setting.
- Some of the questions may be personal in nature. I have the right to stop participating at any time or have my data destroyed even after I have responded.
- While it is not possible to identify all potential risks in an experimental procedure, I understand that an effort has been made by the researcher(s) to minimize them.
- I can remove myself from this study at any time by contacting `Harold Belleview`, `hb@gstate.edu`.
- Intended purpose of study: `Determine beliefs of undergraduate students at Grand State University concerning evolution.`

Name: _____ Signature: _____ Date: _____

Consent to Serve as a Subject in Research   Survey #___
I am at least 18 years of age and consent to serve as a subject in this survey.

- I understand the purpose of this survey and have had opportunities to ask questions.
- My survey responses will be kept confidential and responses will not be publicly disseminated outside of a class setting.
- Some of the questions may be personal in nature. I have the right to stop participating at any time or have my data destroyed even after I have responded.
- While it is not possible to identify all potential risks in an experimental procedure, I understand that an effort has been made by the researcher(s) to minimize them.
- I can remove myself from this study at any time by contacting _____
- Intended purpose of study: _____

Name: _____

Signature: _____ Date: _____

Some activities involving human subjects do not need to pass through institutional review. In colleges and schools, the regular testing of students is not covered by the research principles for human subjects, although many students probably wish it was. Student projects such as surveys in a class setting that are approved by the instructor do not have to go through individual review at the university level. However the instructor needs to request permission to conduct the class projects, and the results of the student projects can only be used within the class and cannot be published. To publish them requires additional review. The surveys should not gather information that could be used to incriminate the respondents.

Even for student projects of brief duration, the principles and practices governing human subjects must all be obeyed faithfully. Here are some requirements

1. Before beginning any project involving human subjects, you must complete training on the treatment of human subjects, and supply a certificate proving it to your instructor. Your college may have a web site that provides training and a certificate; otherwise, go to
   `http://phrp.nihtraining.com/users/login.php` .
2. If you are planning an experiment involving your own health, you must restrict the investigation to measuring normal activities. If you normally drink milk in the morning, it would be acceptable to measure your temperature before and after drinking it to see if your temperature changes. It would not be acceptable to examine how much milk you can drink before you throw up. If there is any possibility the experiment you are proposing might threaten your health, you will need a signed certificate from a doctor stating that there are no health risks. Your instructors may choose not to approve such experiments in any event, and you must consult them in advance. Thus, even if you ordinarily smoke, you cannot conduct an experiment where you look at the effect of smoking on your heart rate, because a doctor cannot sign a statement that smoking carries no health risks.
3. If you are planning an experiment involving the health of other individuals, similar considerations apply. You must restrict the investigation to measuring normal daily activities. In addition, you must obtain a signed consent form from every individual you investigate. Any procedure that could possibly affect health or welfare will require a signed doctor's certificate for each individual. Your instructors may choose not to approve such experiments in any event, and you must consult them in advance.
4. Children (minors) are not legally able to provide consent, and their participation in any experiment is only possible with signed permission from parents or guardians. In most regions of the United States, those 18 years and older are legally adults, while those younger are children, but you should check that these guidelines apply to you.

5. If you are administering a test or conducting a survey, there are several requirements.

   a. Every participant must enter into the process voluntarily, must be informed of the purpose of the research, and must be given the right to withdraw both before and after the survey is conducted.

   b. The anonymity of every participant must be preserved.

For answers to many other questions, such as the propriety of compensating research subjects, see HHS (2010). Carefully respecting the rights of human subjects certainly adds to the length and difficulty of research on people, but these procedures are unquestioned and standard in today's scientific community.

Assignments

2.1 **Laboratory inquiry**

   Purpose Create a quantitative data set in a controlled laboratory setting and subject it to statistical analysis.

   Background This assignment takes place in a formal laboratory setting. Your instructors will tell you what scientific equipment you have available to select from. The questions you can ask will be constrained by the equipment. However, your ability to collect quantitative data should improve because you will be able to work with instruments not typically available at home. You may not be as inventive as you can be when pursuing questions that make you curious at home, but you should be able to perform more precise work.

   You should perform an investigation chosen from the first and third types listed in Chapter 1; investigate an hypothesis, as in Section 1.3.1, or measure a function as in Section 1.3.3. Plan to use statistics in evaluating your results, and take your data with the idea in mind that you will subject them to statistics later on.

   After a first laboratory session in which you explore equipment, you should turn in a brief description of the experiment you intend to perform. This proposal will give instructors the chance to provide some preliminary advice on the direction you are taking. The more detailed your description, the more they will be able to help you. For a description of proposals, see Section 5.2.

   Length 3–5 typed pages. It is generally not a problem to exceed this length.

   Report Please include the following sections:

   a. *Title.*

   b. *Abstract.* This paragraph should explain the purpose of your inquiry, and then summarize the main results. It should be written in present or past tense. Write this section after you have completed all the work. Imagine that you are trying to prepare a single paragraph that will be published by itself in a newspaper to explain to huge numbers of people what you have found.

   c. *Introduction.* This section should explain the motivation for your inquiry, and should incorporate background information including theories and models. An introduction explains the significance of this inquiry and why you decided to spend time investigating it. It also provides enough essential background information that the reader will be able to understand research in the field.

d. *Experimental design.* Your goal in this section is to enable other people to reproduce your experiment. Therefore, you will need to include a description of all the materials you used, and diagrams that explain how the apparatus was constructed. It is perfectly appropriate for you to explain wrong steps that you took, so as to warn others away from repeating your mistakes. You should use only past or present tense in this section.

e. *Analysis*, including an appropriate statistical treatment of the data and explanation of whether your findings are statistically significant.

f. *Conclusions.* State in this section what you found in your experiment and what you have learned. You should set yourself the goal of being honest. On the one hand, you should not minimize the effort you have put into the experiment, and you should not dismiss or underestimate your own results. On the other hand, you should not claim to have found things that your results do not support. You are free to write about how you might do the experiment otherwise if you could do it again, and to make suggestions for yourself or others to pursue in the future.

g. *Data.*You should include enough raw data to enable evaluators to check your results. You can either include the data in the body of the report, or else in an appendix. In some cases, you may choose to provide the instructors with electronic copies of your data.

Grading This inquiry will be evaluated according to the criteria in the Inquiry Grading Rubric, Appendix D.

2.2 **Inquiry peer review**

Background Your instructor will supply you with the report written by one of your colleagues in the class. You should read it, and evaluate it according to the Inquiry Grading Rubric in Appendix D. A checklist for you to fill out is available on the course website.

What to turn in You should fill out the checklist assigning points to the project you have read, and turn in the checklist. However, even more important is that you write comments on the report that you could give to the author to help him improve it. You should type up these comments and submit them. They should be a minimum of half a page. If you want to make additional written comments on your copy of the project and turn that in as well, you may.

2.3 **Human subjects training**

If you have previously obtained a certificate showing that you have conducted training on human subjects in research, you only need to turn it in. Otherwise, go to a website indicated by your instructor, such as

`http://phrp.nihtraining.com/users/login.php`

and complete the training. Make two copies of your certificate. Keep one for yourself – you may well need it in the future – and turn in one.

Note that completing this training process usually takes two hours. It is not a simple form that you fill in and complete in a few minutes.

References

Belmont (1974), http://ohsr.od.nih.gov/guidelines/belmont.html, Retrieved March 2010.

G. Calori, A. D'Angelo, P. Della Valle, G. Ruotolo, L. Ferini-Strambi, C. Giusti, A. Errera, and G. Gallus (1996), The effect of cigarette-smoking on cardiovascular risk factors: a study of monozygotic twins discordant for smoking, *Thrombosis & Haemostasis*, **75**, 14.

E. R. Cohen and J. W. M. DuMond (1965), Our knowledge of the fundamental constants of physics and chemistry in 1965, *Reviews of Modern Physicis*, **37**, 537.

R. M. Douglas and H. Hemilä (2005), Vitamin C for preventing and treating the common cold, *PLoS Medicine*, **2**, 168.

G. J. Gallup (2000), The first measured century, http://www.pbs.org/fmc/interviews/ggallup.htm

J. Kaprio and M. Koskenvuo (1989), Twins, smoking and mortality: a 12-year prospective study of smoking-discordant twin pairs, *Social Science & Medicine*, **29**, 1083.

T. R. Karlowski, T. C. Chalmers, L. D. Frenkel, A. Z. Kapikian, T. L. Lewis, and J. M. Lynch (1975), Ascorbic acid for the common cold. A prophylactic and therapeutic trial, *JAMA*, **231**, 1038.

J. R. Klein and A. Roodman (2005), Blind analysis in nuclear and particle physics, *Annual Review of Nuclear and Particle Science*, **55**, 141.

NIST (2008), http://physics.nist.gov/cgi-bin/cuu/value?e

L. Pauling (1971), The significance of the evidence about ascorbic acid and the common cold, *Proceedings of the National Academy of Sciences, USA*, **68**, 2678.

L. Pauling (1986), *How to Live Longer and Feel Better*, Avon Books, New York.

O. Pfungst (1965), *Clever Hans, the Horse of Mr Von Osten*, Holt, Rinehart and Winston, New York.

P. Singer (1975), *Animal Liberation: A New Ethics for our Treatment of Animals, New York*, Random House, New York.

A. P. R. Wadlund (1928), Absolute X-ray wave-length measurements, *Physical Review*, **32**, 841.

3

Statistics

3.1 Motivations for statistics

Statistics is the mathematics of measurement and data. The data might come from measuring the height of a friend, the sizes of fish in a lake, flipping coins, the rate electrons are ejected from a metal illuminated by light, or mathematics test scores of students in Iowa. The tools of statistics are needed whenever the data are partly predictable and partly influenced by factors too complex fully to understand. Statistics makes it possible to summarize the meaning of huge numbers of measurements in a compact form, and to explain the significance of the measurements even when chance makes them uncertain. Two of the main reasons to learn statistics are to deal with *measurement error* and to be able to describe *distributed quantities*.

Measurement error

Chapter 2 discussed the fact that almost any attempt to measure values obtains numbers that vary from one trial to the next. It provided recipes for dealing with the variation but no explanation for why the recipes work. This chapter returns to the problem and explains the solutions in more detail.

Suppose you are trying to measure the height of your friend Anne with a ruler. You make the measurement a first time and find she is 1.721 m tall. When you repeat the measurement, the meter stick tilts at a slight angle, and you get 1.722 m. You try again, and she slouches slightly, so you get 1.718 m. No matter how often you make the measurement, you continue to get slightly different numbers. The general name for errors of this sort is *measurement error*. These errors are inevitable and maybe a bit annoying. Anne, you think, has a height, and you would like to find it. Statistics will tell you how to assemble the many measurements into one number. Nothing will tell you Anne's true height, for if repeated measurements do not agree completely, the true height comes to depend upon definition as much as measurement. Yet statistics teaches that it is not pointless to make repeated measurements, and it describes how the precision in knowing Anne's height increases the more measurements one takes.

There are some cases where measurements give perfect accuracy. Suppose there are two chairs in a room. Then you can measure the number of chairs as often as you please, and you will always get two. But even measuring integers can be challenging. Elections are measurements of the numbers of people who want to vote for candidates or propositions. When two sides are sufficiently close, errors in recording and transmitting votes can make it nearly impossible to determine who really won, as for the presidential election in Florida in 2000. According to the Federal Election Commission (2000), George W. Bush had 2912,790 votes and Al Gore had 2912,253, but both Democrats and Republicans point to improper actions taken during the election that may have swayed the vote by much larger amounts than the final differences between the two candidates.

Distributed quantities

Statistics is also essential to deal with measurements of distributed quantities, items that fundamentally cannot completely be described by a single number. Even something as apparently simple as the height of a single person such as Anne is not really just one number. It may differ by more than a centimeter from morning to night, and will probably vary by more than that during her lifetime. Many numbers are required for a complete description of her height.

The need for many numbers is even more clear if you want to know the height of all women in the United States. This information is contained in a sequence of around 153,000,000 numbers, constantly changing as women enter and leave the country, are born and die. This sequence of numbers has never fully been recorded and never will be. Gathering it would be enormously expensive, and there is no point since such a large sequence of numbers overwhelms the human mind. We cannot make sense of it. Current understanding of the height of women in the U.S. is based upon a sample of around 4300 women over the age of 20 whose height was measured between 1999 and 2002 (Ogden *et al.*, 2004). But even a sequence of 4300 numbers is impossible to comprehend. So researchers for the National Health and Nutrition Examination Surveys report simple summaries, such as that the average height of U.S. women over 20 years old is 5 feet, 3.8 inches, or 1.62 meters. The *average* is an example of a *descriptive statistic:* a number designed to provide insight into a sequence of numbers. Statistics provides many different ways to present large sequences of numbers in compact forms, and these different ways of representing data help answer questions people would like to answer.

It is natural to wonder whether a sample of 4300 women can possibly provide accurate information about 150,000,000 women. The height of only one woman in 35,000 is being measured. Statistics provides answers to this sort of question too, and describes the level of uncertainty in an average that comes from taking a relatively small random sample.

Value of statistics The word *statistics* was introduced into English in the 1790s by Sir John Sinclair who adopted it from usage in Germany that in turn runs back to a Latin phrase for council of state. He says

Many people were at first surprised at my using the words "statistical" and "statistics", as it was supposed that some in our own language might have expressed the same meaning. But in the course of a very extensive tour through the northern parts of Europe, which I happened to take in 1786, I found that in Germany they were engaged in a species of political enquiry to which they had given the name "statistics," and though I apply a different meaning to that word – for by "statistical" is meant in Germany an inquiry for the purposes of ascertaining the political strength of a country or questions respecting matters of state – whereas the idea I annex to the term is an inquiry into the state of a country, for the purpose of ascertaining the quantum of happiness enjoyed by its inhabitants, and the means of its future improvement; but as I thought that a new word might attract more public attention, I resolved on adopting it, and I hope it is now completely naturalized and incorporated with our language. (Sinclair, 1792)

Thus statistics has been associated since its introduction two centuries ago with the operation of governments, and governments are among the most important entities that gather and interpret data. Two examples of U.S. government data collection have just been mentioned: the Federal Election Commission, and the Centers for Disease Control, which administers the National Health and Nutrition Examination Surveys. There are many others, including the Bureau of Labor Statistics, the Census Bureau, and the National Center for Educational Statistics. The numbers collected by these agencies enable operation of the electoral system and the functioning of the economy and educational systems. Use of statistics spreads far beyond government agencies. Every company relies upon them, as does almost every scientific research project. Mentioning government data as an example helps to underscore a point, which is that statistics arose in response to questions that often have life and death significance.

This significance is far too easy to forget when one begins to plunge into the mathematical theory. So far as mathematics is concerned, statistics provides a set of operations that can be performed on any sequence of numbers, or on more abstract entities called probability distributions. If you take some sequence of numbers and choose a random statistical test, the test will always spit out some number. But beware, as the number can easily be meaningless. This comment applies even to familiar operations like averaging. For example, one could define the Carlyle Index to be the average of a person's age in years, the length of all the hairs on their head laid end to end in meters, and the area of their eyelids in square millimeters. The operation of averaging would give a number. But it would be meaningless because the quantities being averaged are not in any sense the same, and the point of averaging is to find a way to characterize many quantities that vary, but have something basic in common.

The most challenging aspect of statistics is to find the correspondence between questions you want to ask about the world and the complicated and sometimes puzzling mathematical operations the theory makes available. Like other branches of science and mathematics, statistics can grab hold of apparently ordinary English words and give them new meanings. Two words to watch for are *significance* and *confidence*. In common speech, something is significant if it is important, maybe surprising, and has value. In statistics, two averages are significantly different if chance alone is unlikely to have produced the difference. Whether the difference is important or valuable is a separate question. So children two years old are significantly shorter than children six years old, but no one is going to win any prizes by conducting a study to prove it. The difference in height is significant in the statistical sense, but not in the ordinary sense of being surprising and valuable. Conversely, suppose you invented a fluid called Enaprize costing only 30 cents a gallon that acted just like gasoline when you put it in a car. Statistics would say that Enaprize and gasoline were not significantly different in generating power in a car. Finding the absence of a significant difference in this case would be very important and very significant in the ordinary sense. "Confidence" commonly refers to an internal sense of being correct. In statistics, it refers to a range of values around some measured quantity where the true value might fall. The larger the confidence interval around a measured number, the less confident one actually is that the measured value is correct.

Such difficulties, plus the common opinion that statistics can always be twisted to provide the apparent proof of anything desired, conspire to give it an unpleasant reputation. There is a famous saying that there are 'lies, damned lies, and statistics'.[1] This reputation is really not deserved. Using mathematical tools to deal with uncertainty is vastly preferable to blind guessing, and when used with honesty and competence, statistical tools give honest answers.

Chance Statistics is a mathematical response to chance. But what is chance, and why not just get rid of it? Chance describes events that are unpredictable. At the deepest level, the laws of quantum mechanics say that the precise trajectories and locations of particles cannot be known with certainty, and one can only speak of these quantities in terms of likelihood. For most happenings in daily life the

---

[1] Peter Lee, historian of statistics at the University of York has tracked this down to Cornelia Crosse, *The Living Age* 195 (Issue 2523) (1892 Nov. 5), 372–383 at page 379, and claims that common attribution of the saying to Disraeli is due to Mark Twain who made the attribution up. See http://www.york.ac.uk/depts/maths/histstat/lies.htm.

uncertainties due to quantum mechanics are so small as to be completely irrelevant. Other sorts of sources of uncertainty are more important. Knowledge of the physical laws that govern the motion of heat, air, water, and rock is excellent, and should make it possible to predict the weather with great confidence. Weather prediction improved continually over the last 50 years, but everyone is familiar with mistakes in the daily forecast. Weather forecasts a week in advance are terrible. The problem is that making predictions about the weather depends upon knowing the current state of the weather with great accuracy. This would mean knowing the temperature, pressure, and water and wind speed of every bit of earth, water and air from below the Earth's surface to the top of the atmosphere, with at least one set of measurements, say, in every cubic meter of material. Data at this level of resolution have never been available, nor do we yet have computers large enough to process such quantities of data. To make matters worse, chaos theory says that uncertainties about the motion or temperature of air at any point on the planet can grow exponentially in time. So even if you had measurements of temperature and pressure in every cubic meter of air, ignorance about what was happening in every cubic centimeter would eventually doom the predictions. This phenomenon is called the *butterfly effect,* after meteorologist Edward Lorenz, who gave a talk in 1972 entitled *Does the flap of a butterfly's wings in Brazil set off a tornado in Texas?* The answer to the question is *Maybe yes!*

Biology is so complicated that chance plays a particularly large role. Physics has equations that in principle could predict the weather, but biology does not have comparable equations to predict, say, how tall a given person will grow. If you had tissue samples of your friend Anne as an infant, you might vaguely think about analyzing her DNA and predicting her potential for growth, but there are no equations to do so based on fundamental physical law.[2] All the unseen factors, the chocolate cake on December 4, the fall during a camping trip in June, that affect Anne's height as she grows, are grouped together and called chance.

3.2 Reducing many numbers to few

3.2.1 Means and averages

One purpose of the *mean* or *average* is to represent a large collection of numbers with a single number. Averages are especially important in science because when the large collection of numbers is a set of measurements that vary randomly around

---

[2] Which is not to say that predicting height is impossible. One can obtain a height prediction within an inch or so using a child's height and weight and the heights of both parents; see Khamis and Roche (1994), Khamis (2008) , or Tanner *et al.* (2001).

a true value, averaging the measurements together gives a more accurate result than can be obtained from the individual measurements.

To calculate an average, take a collection of N numbers, sum them together, and divide by N. Suppose the numbers are

$$x_1, \ x_2, \ x_3, \ \ldots, x_N.$$

Then the mean value $\bar{x}$ or average of the numbers $x_1, \ldots, x_N$ is

$$\bar{x} = \frac{x_1 + x_2 + x_3 + \cdots + x_N}{N} = \frac{\sum_{i=1}^{N} x_i}{N}. \tag{3.1}$$

The second expression introduces *summation notation*

$$\sum_{i=1}^{N} x_i = x_1 + x_2 + \cdots + x_N. \tag{3.2}$$

Here are some properties of the average that make it useful.

- If all the numbers being averaged together are the same, then the average returns this number. This property is easy to verify because if all numbers $x_1, \ldots, x_N$ are the same number x then

$$\bar{x} = \frac{x_1 + x_2 + \cdots + x_N}{N} = \frac{Nx}{N} = x. \tag{3.3}$$

- The average is always more than the smallest of the numbers in $x_1, \ldots, x_N$. To show why this claim is correct, suppose it is false. If it were false, then one would have

$$\bar{x} < x_1; \quad \bar{x} < x_2; \quad \bar{x} < x_3; \ldots \bar{x} < x_N. \tag{3.4}$$

Add up all the N inequalities in (3.4). Since the left-hand side of each inequality is less than the right-hand side of each inequality, the sum of the left-hand sides must also be less than the sum of the right-hand sides. The result is

$$N\bar{x} < x_1 + x_2 + \cdots + x_N \Rightarrow \bar{x} < \frac{\sum_{i=1}^{N} x_i}{N} \Rightarrow \bar{x} < \bar{x}. \tag{3.5}$$

But the mean $\bar{x}$ cannot be less than itself. The only way one can reach a false mathematical result (without actually making a mistake) is to start with a false assumption. The starting assumption that the average $\bar{x}$ can be less than all the numbers that make it up must have been false.

- The average is always less than the largest of the numbers $x_1, \ldots, x_N$. The demonstration of this claim is very similar to the previous one, and is left as an exercise.

The last two properties of the average can be summarized in words with the statement, *If you have a collection of numbers $x_1, \ldots, x_N$, then the average of these numbers lies somewhere in the middle.* This property of the average is one reason why it gives a useful way to represent the collection of numbers.

Example: *Averaging*

Suppose you have the numbers $x_1 = 1;\ x_2 = 3;\ x_3 = 2;\ x_4 = 2;\ x_5 = 6.$
The average of these numbers is

$$\bar{x} = \frac{1+3+2+2+6}{5} = \frac{14}{5} = 2.8.$$

3.2.2 Median

Sometimes averages do not provide a good way to characterize many numbers. This can happen when a small subset of the numbers is very different from the others. Suppose there are 100 persons in a room. Most of them do not own any airplanes, but one of them founded an aircraft company and owns 100 airplanes. So on average everyone in the room owns one airplane. This statement is true, but very misleading, since a typical person in the room owns no airplanes at all. Or take the salaries of athletes. Their average annual salary in 2004 was $82,540. This average reflects the fact that a small number of athletes in professional teams earn millions per year. However, they are not very representative of the 12,000 people employed as athletes. In fact, half of all athletes earn less than $46,000 a year. This is the definition of the *median:* if you have a collection of numbers, the median divides the upper half of the numbers from the lower half. For the room with 100 people and 100 airplanes, the median number of airplanes people own is zero. When the number of numbers is odd and arranged in increasing order, it is the number in the middle; if the number of numbers is even, take the average of the two numbers in the middle. The median provides a better way to find the typical value of a collection of numbers than the average, particularly if the distribution of numbers is very skewed.

Example: *Median*

Suppose you have the numbers

$$x_1 = 1;\quad x_2 = 2;\quad x_3 = 2;\quad x_4 = 3;\quad x_5 = 6.$$

The median of these numbers is 2, not very far from the mean of 2.8. But suppose instead you have

$$x_1 = 1;\quad x_2 = 2;\quad x_3 = 2;\quad x_4 = 3;\quad x_5 = 6;\quad x_6 = 1000,000.$$

Now the median is 2.5. The average would be 166,669.

3.2.3 Histograms

To find the median of a collection of numbers, divide the numbers into two halves. But why stop there? You could divide them into thirds, fourths, or any other number of groups. *Histograms* provide a systematic way to go about it. To construct a histogram, you first need to select a set of *bins*, and then find how many of your measurements fit into each of the bins. A bin is a range of values you decide to group together.

The idea is best explained with an example. Return to the data for lengths of fish spines in Lakes Dugout and Ormond (Table 2.3). One way to group the measurements is to set up four bins: Bin #1 = 3–4 mm, Bin #2 = 4–5 mm, Bin #3 = 5–6 mm, Bin #4 = 6–7 mm. It is now just a tedious matter of running through the data and counting up exactly how many fish fall in each bin. The numbers are recorded in Table 3.1.

This histogram is quite coarse, as shown in Figure 3.1(A). So many entries are grouped together that one cannot learn much about the shape of the data. A more revealing histogram is obtained by using 20 bins instead: $3.0 \leq l < 3.2$, $3.2 \leq l < 3.4, \ldots, 6.8 \leq l < 7$. The result is plotted in Figure 3.1(B).

Optimizing histograms

If bins are very large, there are not enough points on the horizontal axis to flesh out the shape of the curve. If bins become smaller than the smallest difference between any two distinct points, there may be only one or a few entries per bin, and the histogram becomes flat and shapeless. For the fish spine lengths in Lake Dugout, bins of width 0.01 would produce such a result. A histogram where the number of bins is about the same as the maximum number of entries in any given bin strikes a good balance between these two extremes. To get a quick estimate for this number of bins, take the total number of data points, take the square root, and multiply by 2. For Lake Dugout, there are 128 fish in the sample, and $2 \times \sqrt{128} = 22.6$. So 20 bins produces a good histogram.

Automatic computation of histograms

Since adding up numbers of data points in each bin is tedious, you will usually want a computer to do the work for you. Spreadsheet programs can produce histograms automatically, as described in Appendix A.3.1.

Table 3.1 Histogram of lengths of fish in Lake Dugout with bins of size 1 mm.

Bins for fish spine lengths l in mm	$3 \leq l < 4$	$4 \leq l < 5$	$5 \leq l < 6$	$6 \leq l < 7$
Number of fish in bin	15	75	36	1

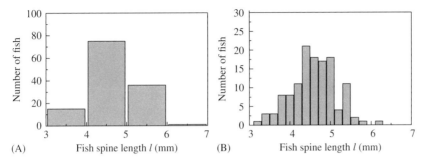

Figure 3.1 (A) Histogram of lengths of fish spines in Lake Dugout, with bins of size 1 mm. (B) Histogram of lengths of fish spines in Lake Dugout, with bins of size 0.2 mm. As the bins become smaller, the number of fish in each bin decreases, and the shape of the curve is fleshed out better. If the bins were to become too small, there would be only 0 or 1 fish in each bin, and the shape of the curve would be lost again.

3.2.4 Standard deviation

The mean or median provide a single number to represent a large collection of data. Suppose one wants to improve characterization of the data with just one more number. What should come next?

After the first number, which records where the center of the data lie, the second number records how widely the data are spread out. It characterizes the *width* of data such as the histograms in Figure 3.1. There is a conventional way to describe the width of a histogram. It provides a *standard* way to capture the typical amount by which a *sample* of data *deviates* from the mean, and is called the *sample standard deviation*.

To get a feel for the standard deviation, return to the data in Table 2.2 for the lubricated ramp and try to determine by eye the typical distance between data points and the mean of 0.324. There are points up to around 0.02 or 0.03 above the mean and 0.02 and 0.03 below, so the typical range of values in the data is around 0.02. There are so many data points in Table 2.3 that it is harder to make such an estimate by scanning the data. Inspecting the histogram in Figure 3.1, however, it seems reasonable to say that the characteristic deviation of the data from the mean is around 1–2 mm above and below.

The formal procedure for finding the standard deviation produces results that are consistent with these rough estimates. A problem that must be overcome is that some data points lie above the mean and others lie below. So if you subtract the mean from data points, sometimes you got positive results and sometimes negative. Adding them up, the positive and negative values cancel out giving zero. The eyeball estimates of standard deviation look at the *distance* between data points and the mean, whether the points lie above or below. So the idea is to take each data point, subtract the mean from it, and *square* the result. Add the squares together. All entries in that sum will be positive and they cannot cancel each other out. Here is the procedure:

- Denote the number of data points, or *sample size*, by N.
- Subtract the mean from each data point in turn.
- Square the result so that it is positive; note that if the quantities one is measuring have units, those units now are squared.
- Add up all the results.
- Divide by sample size minus one, $N - 1$.
- Take the square root, so that the units of the result are the same as the units with which one began.
- The result is called the *sample standard deviation*, s.

More formally, suppose you have N data points, $x_1, x_2, x_3, \ldots, x_N$. First compute the sample mean $\bar{x} = \sum_{i=1}^{N} x_i / N$. From it compute the sample *variance* by

$$s^2 = \frac{\sum_{i=1}^{N} (x_i - \bar{x})^2}{N - 1} \tag{3.6}$$

and the sample standard deviation by

$$s = \sqrt{\frac{\sum_{i=1}^{N} (x_i - \bar{x})^2}{N - 1}}. \tag{3.7}$$

Why divide by $N - 1$ and not N? There is a formal explanation in Section 3.8.3, but one can more simply think of it like this. Suppose one has only one data point, x_1. This data point is the same as the mean; since there is only one point, the data have no width or spread and the standard deviation is undefined. Now suppose one has two data points, x_1 and x_2. There are two pieces of information. One of the pieces of information is used to determine the mean, and one is left over to give some information about the width of the distribution, so divide by 1. If there are three data points, one piece of information is again used up determining the mean, and two are left over to estimate the width, so divide by 2. And so on. Why take the square root at the end? In the process of squaring all the differences between data points and the sample mean, the characteristic size has been changed. For example, if the characteristic distance from data points to the mean is 0.1, after squaring, the characteristic value has become 0.01. This change is undone at the end of the calculation by taking the square root.

Example: *Standard deviation*
Consider again the numbers $x_1 = 1$; $x_2 = 3$; $x_3 = 2$; $x_4 = 2$; $x_5 = 6$ for which the average is $\bar{x} = 2.8$. The sample standard deviation is

$$s = \sqrt{\frac{\left(\begin{array}{c}(1-2.8)^2 + (3-2.8)^2 + (2-2.8)^2 \\ +(2-12.45)^2 + (6-2.8)^2\end{array}\right)}{(5-1)}} = \sqrt{11.84/4} = 1.72.$$

Spreadsheets can compute standard deviations. If data are stored in cells A1,..., A5, then their standard deviation is =STDEV(A1:A5).

3.3 Probability distributions

3.3.1 Definitions

Probability distributions are one of the central ideas underlying statistics. They describe situations where a measurement can be repeated many, many times, where the measurements produce a range of different values, and where successive measurements are completely independent of one another. Every probability distribution has a *sample space* The sample space is a complete list of measurements or events that happen with some probability greater than zero. The sample space can be continuous, or it can be discrete. For a ball rolling down a track, the sample space is the continuous range of times that the ball might take to make the trip. For fish in two lakes, the sample space is the continuous set of lengths between 3 and 7 mm that the spines can have. For a coin being flipped, the sample space is the two values, heads = 1, tails = 0, that the coin can take.

A probability distribution is a function that takes a value for every point X in the sample space. It describes how likely each measurement value X is to appear. Notation used in this chapter for probability spaces appears in Table 3.2.

3.3.2 Discrete probability distributions

Probability distributions are easiest to define when the state space is discrete, and the only possible measurements come from a finite set of values $\{x_1, \ldots, x_M\}$. Imagine performing a very large number $N \gg M$ of measurements. Suppose that the value x_1 appears N_1 times, and x_2 appears N_2 times, and so on up to x_M which appears N_M times. Adding up the times each particular measurement value appears gives the total number of measurements:

Table 3.2 Conventions for notation in this chapter.

x_i or $x_1, x_2, \ldots, x_N$	Values you obtain in a series of N measurements that all aim to measure the same thing. N is the sample size.
$\mathbf{x}_j$ or $\{\mathbf{x}_1, \mathbf{x}_2, \ldots, \mathbf{x}_M\}$	For measurements with a discrete set of outcomes, a complete listing of all M possible outcomes; the sample space.
X	A variable that ranges over allowed values in a discrete or continuous sample space.
$X_1, X_2, \ldots, X_N$	A sequence of variables each of which ranges over allowed values in a sample space, but which will be used to describe a sequence of N measurements.
p	The probability of a single measurement, or of a range or collection of measurements.
$\mathcal{P}(X)$	A probability density, whose integrals over X give probabilities p.

$$N = \sum_{j=1}^{M} N_j. \tag{3.8}$$

Define the probability distribution p to be the fraction of times each measurement occurs:

$$p(\mathbf{x}_j) \equiv N_j/N. \tag{3.9}$$

From Equation (3.8) it follows that

$$\sum_{j=1}^{M} p(\mathbf{x}_j) = 1. \tag{3.10}$$

Alternative notation for this same equation is

$$\sum_X p(X) = 1, \tag{3.11}$$

since $\{\mathbf{x}_1, \ldots, \mathbf{x}_M\}$ is the state space of all values that X can take.

Sums of probabilities

The number of times that measurement $\mathbf{x}_1$ *or* measurement $\mathbf{x}_2$ shows up is $N_1 + N_2$. The fraction of times, or probability, that these measurements will appear is $(N_1 + N_2)/N = p_1 + p_2$. In general, adding together probabilities $p_1, p_2, p_3, \ldots$ gives the total probability that any one of these events occurs during one measurement. The meaning of Equation (3.10) is that every time one makes a measurement, it is certain that *some* possible value will appear.

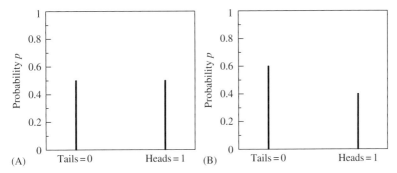

Figure 3.2 (A) Probability p of finding heads and tails for a fair coin. (B) Probability p of finding heads and tails for a coin where tails come up 60% of the time.

Single coin flip

The simplest example of a discrete probability distribution comes from flipping a coin. The sample space is Tails (0), Heads (1). Assigning 0 and 1 to heads and tails is an arbitrary choice; any two integers would do. If the coin is fair, then heads and tails have equal probability, so $p(0) = \frac{1}{2}$ and $p(1) = \frac{1}{2}$. As required by Equation (3.10), $p(0) + p(1) = 1$. This distribution is plotted in Figure 3.2(A). There are no error bars on the probability distribution. It is mathematically perfect, exact. Any measurements of real coins should be reported with error bars, but that is because they are real objects, not mathematical abstractions. Figure 3.2(B) shows the probability distribution for a weighted coin where tails come up 60% of the time. Now $p(0) = 0.6$, $p(1) = 0.4$, and once again $p(0) + p(1) = 1$. This distribution could represent many things other than a weighted coin. It could represent the fraction of a voting population that supports building a new toll road. It could represent the fraction of headache sufferers who respond more quickly to RCS Aspirin. The probability distribution is supposed to represent the real, true answer one would find if one could make completely exhaustive sets of measurements.

Sequences of measurements: Products of probabilities

So far, flipping a coin once was regarded as one measurement. But it is also possible to regard flipping a coin twice, or flipping two identical coins once, as one measurement.

What are the probabilities associated with flipping a coin twice? Figure 3.3(A) shows the probability of every possible outcome. There are four of them, and each one is equally likely, with probability $\frac{1}{4}$.

In general the probability of making measurements $\mathbf{x}_i$ and $\mathbf{x}_j$ in sequence is the product $p(\mathbf{x}_i)p(\mathbf{x}_j)$. The sample space is every possible combination $\mathbf{x}_i\mathbf{x}_j$ and when there are M possible values for $\mathbf{x}_i$, the sample space for $\mathbf{x}_i\mathbf{x}_j$ has M^2

members. To see how the probability distribution arises, imagine making two independent measurements a total of N^2 times, with $N \gg M$. Value x_1 shows up a fraction $p(x_1)$ times in the first measurement, or $N^2 p(x_1)$ times overall. Out of all the measurements where $p(x_1)$ showed up first, x_2 shows up a fraction $p(x_2)$ of the times, for a total of $N^2 p(x_1) p(x_2)$ times. Dividing through by the N^2 trials gives a probability for $p(x_1, x_2)$ of $p(x_1) p(x_2)$. Similarly, the probability of a precise sequence $X_1, X_2, \ldots, X_l$ is $p(X_1) p(X_2) \ldots p(X_l)$.

Probabilities of sequences, neglecting the order

The probability distribution of a sequence of independent events is very easy to calculate. But it does *not* correspond to the information you will want if you conduct a large number of trials in an experiment.

Suppose you are surveying 10 people. Which question will you want to answer?

Q1 What is the probability that the first person said "Yes" *and then* the second person said "No" *and then* the third person said "No" *and then* the fourth person said "Yes".... *and then* the tenth person said "Yes"?

Q2 What is the probability that four people said "Yes" and six people said "No"?

These two different questions have different sample spaces and different probability distributions. The sample space for the first question is the space of all possible sequences of 10 consecutive "Yes/No" answers. There are $2^{10} = 1024$ different such sequences. The sample space for the second question is the set of integers ranging from 0 to 10, corresponding to 0 "Yes" answers, 1 "Yes" answer, up to 10 "Yes" answers. There is virtually never a case where anyone is interested in the answer to question 1, except as a stepping stone to answering question 2. The number of possible sequences grows so quickly that even writing them all down becomes impossible. In a survey of 100 people, if you tried to make a chart using a space just 1 millimeter wide for every possible sequence, it would take a strip of paper more than 10^{20} times as long as the radius of the Earth. Not practical at all. So, returning to the example of two consecutive coin flips, consider Figure 3.2(B). Instead of showing the probability of every possible sequence of heads and tails, it shows the probability of obtaining 0, 1, and 2 heads. There is one way to obtain 0 heads (TT), two ways to obtain 1 head (HT, TH), and one way to obtain two heads (HH). Therefore the probabilities of 0, 1, and 2 heads are $p(0) = \frac{1}{4}$, $p(1) = \frac{1}{2}$, and $p(2) = \frac{1}{4}$ (Figure 3.3).

Binomial distribution for a fair coin

The *binomial distribution* gives the probability of finding m heads in N throws of a fair coin. There are three ways to compute it.

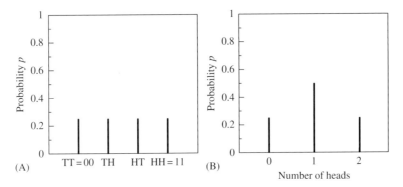

Figure 3.3 (A) Probability p for each outcome in a sequence of two coin flips. T represents tails and H represents heads. (B) Probability p for 0, 1, and 2 heads in a sequence of two coin flips.

The first is to turn to a spreadsheet, which has this distribution built in as a function. To find the probability of flipping a coin 24 times and obtaining 10 heads, just type into a cell =BINOMDIST(10,24,0.5,0)

The = means "compute a function for me." The first argument, 10, means. "I just had 10 heads come up..." The second argument, 24, means "... out of 24 flips." The third argument, 0.5, means "The coin is fair, and has a 50–50 chance of coming up heads." The final argument, 0, means "Tell me the probability of obtaining this number of heads," while if it were 1 it would mean "Tell me the cumulative probability of obtaining this number of heads or less."

Second, lurking behind the spreadsheet's formula is a mathematical theory for binomial probabilities that makes it possible to compute them from simpler mathematical operations. The probabilities are related to the coefficients of a polynomial. To generate all possible sequences of heads and tails in N coin flips, consider the polynomial

$$(H + T)^N. \tag{3.12}$$

For one flip ($N = 1$) the polynomial is $H + T$. For two flips ($N = 2$) it is $(H + T)(H + T) = H^2 + HT + TH + T^2 = H^2 + 2HT + T^2$. For three flips ($N = 3$) it is $H^3 + 3H^2T + 3T^2H + T^3$. Before terms are regrouped, there are 2^N terms in each polynomial, and each term has a probability of $1/2^N$ to arise. After terms are regrouped, there are $N + 1$ terms. The integer coefficient of each term is called the *binomial coefficient,* and to obtain the probability of this many heads and tails, divide the binomial coefficient by the probability of each individual term, namely 2^N. Thus, for a sequence of three coin flips the probability of m heads, $p(m)$, is $p(0) = \frac{1}{8}$, $p(1) = \frac{3}{8}$, $p(2) = \frac{3}{8}$, and $p(3) = \frac{1}{8}$. Mathematicians have found that the binomial coefficient corresponding to m heads and $N - m$ tails is

Table 3.3 Pascal's triangle.

2^N	# heads $m \rightarrow$ 0	1	2	3	4	5	6	7	8	9	10
$2^N \downarrow$	1										
$2^1 = 2$	1	1									
$2^2 = 4$	1	2	1								
$2^3 = 8$	1	3	3	1							
$2^4 = 16$	1	4	6	4	1						
$2^5 = 32$	1	5	**10**	10	5	1					
$2^6 = 64$	1	6	15	20	15	6	1				
$2^7 = 128$	1	7	21	35	35	21	7	1			
$2^8 = 256$	1	8	28	56	70	56	28	8	1		
$2^9 = 512$	1	9	36	84	126	126	84	36	9	1	
$2^{10} = 1024$	1	10	45	120	210	252	210	120	45	10	1

$$\frac{N!}{m!(N-m)!}. \tag{3.13}$$

The symbol ! means to compute the *factorial*, which is

$$N! = N \times (N-1) \times (N-2) \times (N-3) \times \cdots \times 3 \times 2 \times 1. \tag{3.14}$$

So finally, the probability of m heads from N flips is

$$p(m) = \frac{1}{2^N} \frac{N!}{m!(N-m)!}. \tag{3.15}$$

Third, binomial coefficients can be computed from *Pascal's triangle*, which is shown in Table 3.3. To compute the number at some point in a new row, add the two numbers one row above to the left and to the right. For example, the number **10** in bold is the sum of the 4 above to the left, and 6 above to the right. If a number is missing above to the left or the right, treat the missing number as zero. To find the probability of m heads in a sequence of N flips, divide the $(m + 1)$'st entry in a row by 2^N. The row labeled $2^4 = 16$ indicates that in a sequence of four coin flips one should expect all tails one time in 16, one head four times in 16, two heads six times in 16, three heads four times in 16, and four heads one time in 16. The probabilities of obtaining various numbers of heads are obtained, as in Equation (3.15), by dividing through by 2^N.

Binomial distribution for a weighted coin

With a slight generalization, one can find the probability of m heads and $N - m$ tails during N flips of a weighted coin, where the probability of getting heads is some fraction f that lies between 0 and 1. In spreadsheets, the fraction f is entered as the third argument of the function BINOMDIST. Suppose you survey

24 people randomly from a population where 65% favor building a toll road. The probability that 12 of the 24 will say they favor building the road is given by =BINOMDIST(12,24,0.65,0)=0.05. The mathematical expression for the probability is

$$p(m) = f^m (1 - f)^{N-m} \frac{N!}{m!(N-m)!}. \tag{3.16}$$

3.3.3 Continuous probability distributions

Probability distributions are more difficult to define if the values X in the sample space are continuous. It is no longer possible to sum over all the values as in Equation (3.10). Indeed, since the number of possible values is infinite, the odds that any one particular precise value $\mathbf{x}$ will occur during N measurements is vanishingly small. The solution to this problem is to group the measurement values into bins, made as narrow as one likes, and to construct a histogram showing how many times measurements fall into each bin. In particular, take bins of width dX. Label the bins with the integer j, so that each bin extends from

$$\mathbf{x}_j = j\, dX \quad \text{to} \quad \mathbf{x}_{j+1} = (j + 1)\, dX. \tag{3.17}$$

As in the case of discrete distributions, imagine taking N measurements, and let N_j give the number of times that the measurement x falls into bin j. One might expect the continuous probability distribution to be defined as in Equation (3.9) by $p = N_j/N$, which worked for discrete distributions. Alas, this definition creates a problem when the sample space is continuous. The problem is that if one cuts the size of the bin near X in half, the number of measurements N_i that falls into it will drop by a factor of 2 as well. That is, a probability distribution function defined this way will depend on the size of the bins that was used to define it. It is much nicer to set up the probability distribution in such a way that once the bin size dX becomes small enough, the bin size drops out and does not affect the results any more. The way to achieve this goal is to define

$$\mathcal{P}(\mathbf{x}_j) \equiv \frac{N_j}{N\, dX}, \tag{3.18}$$

where $\mathbf{x}_j$ is defined in Equation (3.17). The measurements N_j must still sum up to N, and this means that

$$\sum_j \mathcal{P}(\mathbf{x}_j)\, dX = \sum_X \mathcal{P}(X)\, dX = \sum_j \frac{N_j}{N} = 1. \tag{3.19}$$

Example involving spine lengths

To see how to convert a histogram to a probability distribution, return to the histogram for fish sizes in Lake Dugout shown in Figure 3.1. The total area contained

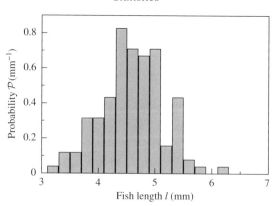

Figure 3.4 Probability distribution for lengths of fish in Lake Dugout, derived from histogram in Figure 3.1. The only thing that has changed is the vertical scale.

within the histogram is 128×0.2mm, since if all the bars are laid end-to-end, their total height must add up to 128 fish, while the width of each bar is 0.2 mm. Now as in Equation (3.18) take the number of measurements in each bin and divide by the total number of measurements and the width of the bin: 128×0.2 mm. Dividing the height of each bar by 25.6 mm is just what is needed to produce a curve of unit area, and the result appears in Figure 3.4.

Continuous limit

In the limit that the bin size dX becomes very small, the left-hand side of Equation (3.19) is nothing but the definition of the integral as a Riemann sum, meaning that

$$\int \mathcal{P}(X)\,dX = 1 : \tag{3.20}$$

the probability distribution $\mathcal{P}(X)$ is a continuous function and the area under it is 1.

In general if one picks an interval $[\mathbf{x}_a, \mathbf{x}_b]$, then the integral

$$p = \int_{\mathbf{x}_a}^{\mathbf{x}_b} \mathcal{P}(X)dX \tag{3.21}$$

gives the fraction of times or probability p that experimental measurements lie between $\mathbf{x}_a$ and $\mathbf{x}_b$. The area under the whole curve is 1 because every measurement is certain to give some value.

Example involving time measurements

Figure 3.5 shows a possible probability distribution for the time measurements in the lubricated channel of Table 2.2. The peak of the distribution is at the measured mean value of 0.324 sec, and the distribution is around 0.018 sec wide. The total

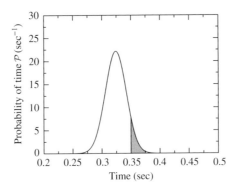

Figure 3.5 Theoretical distribution for the probability of taking various amounts of time to fall down a lubricated ramp. The shaded region shows the area under the curve between 0.35 and 0.375 seconds, and contains 7.2% of the total area.

area under the curve is 1, as it must be for a probability distribution. The shaded region shows the area under the curve between 0.35 and 0.375 seconds. The area is 0.072. This means that the probability p of observing a ball falling with a time that lies between 0.35 and 0.375 seconds is 7.2%. Right at 0.35 seconds, the height of the probability curve is 7.81 sec^{-1}. This does not mean that the probability of finding a ball that takes this much time to fall is 7.81. The value is greater than one and that would make no sense. It also is not dimensionless as probabilities must be. What this value does mean is that if one takes a tiny time interval dt in the neighborhood of 0.35 seconds, then the probability of finding a fall time in the interval between 0.35 seconds and $0.35 + dt$ seconds is 7.81 $\text{sec}^{-1} \times dt$. For this argument to be correct, the time interval dt needs to be small enough that the probability distribution is roughly constant. For example, if one chooses a time interval of 0.001 seconds, then the probability of finding a value between 0.35 and 0.351 seconds is approximately 0.0078, or 0.78%. As the time interval dt shrinks to zero, the probability of finding a fall time in the interval drops to zero, because the probability of obtaining any particular exact value in a measurement with continuous outcomes is vanishingly small.

Main point For continuous probability distributions $\mathcal{P}(X)$, areas under the curve give probabilities. The probability p of obtaining a measurement in the interval between $\mathbf{x}_a$ and $\mathbf{x}_b$ is

$$p = \int_{\mathbf{x}_a}^{\mathbf{x}_b} \mathcal{P}(X)\,dx. \tag{3.22}$$

The value of $\mathcal{P}(X)$ at a point X is not directly meaningful.

Normal distribution

Any positive function with unit area can be a probability distribution. But there is one function above all that is important for probability distributions. It is the *normal distribution,* also known as the *Gaussian distribution* or the *bell curve.* The normal distribution with mean μ and width σ is

$$\mathcal{P}(X) = \frac{1}{\sqrt{2\pi\sigma^2}} e^{-(X-\mu)^2/(2\sigma^2)}. \qquad (3.23)$$

Figure 3.5 shows a picture of a normal distribution with mean $\mu = 0.324$ and width $\sigma = 0.018$. Normal distributions are important because they provide an astonishingly accurate description of the probability of measuring averages. This topic will be discussed further in Section 3.6.

3.4 Connecting data and probability distributions

The central concept of statistics is to make use of mathematical results about probability distributions to estimate the likelihood of observing various results from data. For this process to work, you will have to make a connection between the data you have obtained and some theoretical probability distributions. Sometimes you will take just a few numbers from your data, use them to obtain a probability distribution, and then calculate the probability of other observations. Sometimes you will calculate the probability that different theoretical models could explain your data. In all cases, you need ways to find a connection between your observations and probability distributions.

3.4.1 Sampling

In the lab, you make a set of measurements, for example of the time it takes a ball to roll down a ramp, $x_1, \ldots, x_N$. To think of these data from the point of view of probability distributions, you assume that there is a distribution $\mathcal{P}(X)$ that gives the probability of obtaining your measurements. The N measurements you have taken are a *sample* from this distribution. The relative probabilities of obtaining different values of x_i are given by $\mathcal{P}(X)$. If a different person were to make measurements on the same system, the particular numbers $x_1, \ldots, x_N$ they would find would be different, but they would be statistically equivalent because they came from the same distribution. When you conduct an experiment, you rarely know what the distribution $\mathcal{P}(X)$ is, and need to try to deduce it from your data. In principle, this could be done with certainty by taking an infinite quantity of data and plotting a histogram of the outcome. In practice it cannot be done with certainty, and instead proceeds in an approximate fashion that makes use of quantities such as the mean and standard deviation.

3.4.2 True mean and true standard deviation

Discrete distributions

Two central ingredients connecting data to probability distributions are the sample mean and sample standard deviation of your data. They can be used to make a connection with theoretical distributions, because the theoretical distributions have means, variances, and standard deviations too. To find them, imagine drawing a sample of data from the probability distribution, where the number of observations N goes to infinity. The average of this sample of data is called the *population mean* or *true mean*, μ. The variance of this sample of data is called the *population variance* or *true variance*, σ^2, and its square root is the *population* or *true standard deviation* σ.

For example, consider a fair coin, where tails are represented by 0, and heads by 1. By definition, the coin is fair if when flipped infinitely often, heads and tails come up equally often. This means that when N is large enough, to any desired accuracy $N/2$ of the flips will be tails and $N/2$ will be heads. So the true mean for the coin flip is

$$\mu = \frac{1}{N} \sum_{i=1}^{N} x_i = \frac{1}{N} \left(\frac{N}{2} \times 0 + \frac{N}{2} \times 1 \right) = \tfrac{1}{2}. \tag{3.24}$$

Another way to put it is to say that the probability of getting heads is $p(1) = \tfrac{1}{2}$ and the probability of getting tails is $p(0) = \tfrac{1}{2}$, so the true mean is

$$\mu = p(0) \times 0 + p(1) \times 1 = \tfrac{1}{2} \times 0 + \tfrac{1}{2} \times 1 = \tfrac{1}{2}. \tag{3.25}$$

In general, take any discrete probability distribution where the sample space has values $x_1, x_2, \ldots, x_M$. Then the true mean of the probability distribution is obtained by choosing a sample size N that is very, very, large, and computing

$$\mu = \frac{1}{N} \sum_{i=1}^{N} x_i. \tag{3.26}$$

A fraction $p(x_1)$ of the time the variable x_i comes up with value x_1. A fraction $p(x_2)$ of the time the variable x_i comes up with value x_2. This means there are $Np(x_1)$ terms where x_i equals x_1, $Np(x_2)$ terms where x_i equals x_2, and so on. Therefore Equation (3.26) can be rewritten by taking each value x_i, multiplying by the number of times $Np(x_i)$ it shows up in the sum, and summing over all the x_i. The result is

$$\mu = \frac{1}{N} \sum_j \mathbf{x}_j N p(\mathbf{x}_j) = \sum_X X p(X). \tag{3.27}$$

Example: *Find the true mean for four flips of a fair coin*
The sample space is {0, 1, 2, 3, 4}, corresponding to 0, 1, 2, 3, and 4 heads from the four flips. From Table 3.3 the probabilities of these five outcomes are

$$p(0) = \tfrac{1}{16}, \ p(1) = \tfrac{4}{16}, \ p(2) = \tfrac{6}{16}, \ p(3) = \tfrac{4}{16}, \ p(4) = \tfrac{1}{16}. \tag{3.28}$$

Therefore, the true mean is

$$\mu = 0 \times p(0) + 1 \times p(1) + 2 \times p(2) + 3 \times p(3) + 4 \times p(4) = \tfrac{32}{16} = 2. \tag{3.29}$$

Half the time the coin comes up heads, so the expected value after four flips is 2.

To find the true variance and standard deviation, let N be very large and compute

$$\sigma^2 = \frac{1}{N-1} \sum_{i=1}^{N} (x_i - \mu)^2. \tag{3.30}$$

Because N is very large, the difference between N and $N - 1$ can be neglected. A fraction $p(\mathbf{x}_1)$ of the time the variable x_i comes up with value $\mathbf{x}_1$. A fraction $p(\mathbf{x}_2)$ of the time the variable x_i comes up with value $\mathbf{x}_2$. This means there are $Np(\mathbf{x}_1)$ terms where x_i equals $\mathbf{x}_1$, $Np(\mathbf{x}_2)$ terms where x_i equals $\mathbf{x}_2$, and so on. Therefore Equation (3.30) can be rewritten as

$$\sigma^2 = \frac{1}{N} \sum_j (\mathbf{x}_j - \mu)^2 N p(\mathbf{x}_j) \tag{3.31}$$

$$\Rightarrow \sigma = \sqrt{\sum_X (X - \mu)^2 p(X)} . \tag{3.32}$$

Example: *Find the true standard deviation for one fair coin*
The sample space is {0, 1}, the true mean is $\mu = \tfrac{1}{2}$, and $p(0) = p(1) = \tfrac{1}{2}$. Therefore

$$\sigma = \sqrt{\tfrac{1}{2} \times \left(0 - \tfrac{1}{2}\right)^2 + \tfrac{1}{2} \times \left(1 - \tfrac{1}{2}\right)^2} = \tfrac{1}{2}. \tag{3.33}$$

This result is obvious in retrospect because whether the coin comes up heads or tails, the distance from the result to the true mean is always exactly $\tfrac{1}{2}$.

Example: *Find the true standard deviation for one coin whose probability of coming up heads is* f.

The sample space is $\{0, 1\}$, the true mean is $\mu = f$, $p(0) = 1 - f$, and $p(1) = f$. Therefore

$$\sigma = \sqrt{(1 - f) \times (0 - f)^2 + f \times (1 - f)^2} = \sqrt{f(1 - f)}. \qquad (3.34)$$

This result is very useful for analyzing responses to Yes/No questions, or other cases where a measurement gives two values that show up unequal fractions of the time.

Summary: *For any discrete probability distribution, there are three particularly useful sums:*

$$\sum_X p(X) = 1 \qquad (3.35)$$

$$\sum_X X p(X) = \mu \qquad (3.36)$$

$$\sum_X (X - \mu)^2 p(X) = \sigma^2. \qquad (3.37)$$

Continuous distributions

Now imagine drawing a very large sample N of measurements from a continuous distribution $\mathcal{P}(X)$. Pick a small interval dX, and consider $\mathbf{x}_j = j\, dX$, where j is an integer. The fraction of measurements that lies in the interval between $\mathbf{x}_j = j\, dX$ and $\mathbf{x}_j + dX = \mathbf{x}_{j+1}$ is $\mathcal{P}(\mathbf{x}_j)dX$. So one can use the definition of the Riemann integral to find

$$\mu = \frac{1}{N} \sum_{i=1}^{N} x_i = \frac{1}{N} \sum_{j=-\infty}^{\infty} \mathbf{x}_j \, N\mathcal{P}(\mathbf{x}_j)dX = \sum_X X\mathcal{P}(X)\,dX$$

$$\Rightarrow \mu = \int_{-\infty}^{\infty} X\mathcal{P}(X)dX. \qquad (3.38)$$

Similarly, the true standard deviation is

$$\sigma = \sqrt{\int_{-\infty}^{\infty} (X - \mu)^2 \mathcal{P}(X)\,dX}. \qquad (3.39)$$

Example: *True mean and standard deviation for the normal distribution*
The element of surprise in this example is probably reduced by the fact that two parameters in the normal distribution are μ and σ, and the perceptive reader may have guessed that the reason is that they are the true mean μ and standard deviation σ. Still, provided your calculus is ready for a workout, it is worth checking. The true mean is given by

$$\int X \frac{1}{\sqrt{2\pi\sigma^2}} e^{-(X-\mu)^2/(2\sigma^2)} dX$$

$$= \int (X-\mu) \frac{1}{\sqrt{2\pi\sigma^2}} e^{-(X-\mu)^2/(2\sigma^2)} dX + \int \mu \frac{1}{\sqrt{2\pi\sigma^2}} e^{-(X-\mu)^2/(2\sigma^2)} dX. \quad (3.40)$$

The first integral in (3.40) vanishes because $X-\mu$ is odd and the rest of the integrand is even. The second integral is just μ times the integral of the normal distribution. But the integral of the normal distribution is 1 because it is a probability distribution, so indeed the true mean is μ:

$$\mu = \int X \frac{1}{\sqrt{2\pi\sigma^2}} e^{-(X-\mu)^2/(2\sigma^2)} dX. \quad (3.41)$$

To find the standard deviation, use the following trick. Introduce a variable ϵ into the argument of the exponential and differentiate by it:

$$\int (X-\mu)^2 \frac{1}{\sqrt{2\pi\sigma^2}} e^{-\epsilon(X-\mu)^2/(2\sigma^2)} dX \quad (3.42)$$

$$= \int \frac{1}{\sqrt{2\pi\sigma^2}} (-2\sigma^2) \frac{d}{d\epsilon} e^{-\epsilon(X-\mu)^2/(2\sigma^2)} dX$$

$$= (-2\sigma^2) \frac{d}{d\epsilon} \int \frac{1}{\sqrt{2\pi\sigma^2}} e^{-\epsilon(X-\mu)^2/(2\sigma^2)} dX$$

$$= (-2\sigma^2) \frac{d}{d\epsilon} \frac{1}{\sqrt{\epsilon}} \int \frac{1}{\sqrt{2\pi\sigma^2/\epsilon}} e^{-(X-\mu)^2/(2\sigma^2/\epsilon)} dX. \quad (3.43)$$

The integral in (3.43) is the area under a new normal distribution with σ^2 replaced by σ^2/ϵ. Like all normal distributions, it has unit area so (3.43) becomes

$$-2\sigma^2 \frac{d}{d\epsilon} \frac{1}{\sqrt{\epsilon}} = \frac{\sigma^2}{\epsilon^{3/2}}. \quad (3.44)$$

Equating (3.42) with (3.44) and setting $\epsilon = 1$ gives the variance and standard deviation

$$\sigma^2 = \int (X-\mu)^2 \frac{1}{\sqrt{2\pi\sigma^2}} e^{-(X-\mu)^2/(2\sigma^2)} dX$$

$$\Rightarrow \sigma = \sqrt{\int (X-\mu)^2 \frac{1}{\sqrt{2\pi\sigma^2}} e^{-(X-\mu)^2/(2\sigma^2)} dX}. \quad (3.45)$$

Summary: *For any continuous probability distribution, there are three particularly useful integrals:*

$$\int \mathcal{P}(X)\, dX = 1 \tag{3.46}$$

$$\int X\mathcal{P}(X)\, dX = \mu \tag{3.47}$$

$$\int (X - \mu)^2\, \mathcal{P}(X)\, dX = \sigma^2. \tag{3.48}$$

3.5 What happens to averages as N increases

3.5.1 A drunken sailor

There is a very powerful mathematical result that tells you what happens to sample means as you take more and more data. It shows that random errors can be reduced as much as you want by taking enough data, and tells you just how fast they go away.

Start with Moe, the drunken sailor. He wants to go home, but can't remember which way he is going. Each step he takes is in a completely random direction, to the backward (-1) or to the forward ($+1$). The sample space for Moe is $X \in \{-1, 1\}$; X can take values -1 or 1.

With each step, how far does he go on average? He goes forward with probability $p(1) = \frac{1}{2}$ and backward with probability $p(-1) = \frac{1}{2}$, so on average he goes a distance of zero:

$$\mu = \sum_{X \in \{-1,1\}} Xp(X) = -1 \times \tfrac{1}{2} + 1 \times \tfrac{1}{2} = 0. \tag{3.49}$$

What about his standard deviation from where he started after one step? He goes either forward or back by one step. In either case, he is distance 1 from where he started, so his variance and standard deviation are 1:

$$\sigma^2 = \sum_{X \in \{-1,1\}} (X - 0)^2 p(X) = 1 \times \tfrac{1}{2} + 1 \times \tfrac{1}{2} = 1 \Rightarrow \sigma = 1. \tag{3.50}$$

Now he takes two steps. How far does he go on average with two steps? Each of the two steps X_1 and X_2 can be forward or backward with equal probability, so again on average he goes nowhere. The distance he goes in two steps is $X_1 + X_2$, the probability that X_1 will take some value and X_2 will take a second value is $p(X_1)p(X_2)$, so the average distance two steps carries him is

$$\mu = \sum_{X_1 \in \{-1,1\}} \sum_{X_2 \in \{-1,1\}} (X_1 + X_2) p(X_1) p(X_2). \tag{3.51}$$

The contributions from the two terms in parentheses must be the same, since the first and second steps must contribute the same amount to the mean, so

$$\mu = 2 \left(\sum_{X_1} X_1 p(X_1) \right) \left(\sum_{X_2} p(X_2) \right) = 2 \left(-1 \times \tfrac{1}{2} + 1 \times \tfrac{1}{2} \right) (1) = 0. \tag{3.52}$$

Not only can one talk about the average distance that Moe goes in two steps, one can also talk about the standard deviation D of two steps. What does this mean? It means that since the steps are random, there is uncertainty in how far he goes in his two steps, since he does not always go precisely the mean distance. The standard deviation of Moe's position from 0 is bigger than zero because half the time he goes a distance 2, forward or backward. On the other hand, it must be less than 2, since the other half of the time he takes one step to the backward, one step to the forward, and goes a net distance 0 back to where he started. When he goes distance 2 or -2, the square of the distance he travels is 4; this happens half the time, while the other half of the time he goes distance 0. So the average of the square of the distance he goes is $4/2 = 2$. Taking the square root gives a standard deviation of $\sqrt{2}$ in the distance he travels after two steps. For those who prefer to see a formal calculation that produces this result for the average value of $(X_1 + X_2)^2$, here it is. To simplify the calculations, recall from Equations (3.49) and (3.50) that $\sum_X X \, p(X) = 0$ and $\sum_X X^2 \, p(X) = 1$:

$$\begin{aligned}
D^2 &= \sum_{X_1} \sum_{X_2} (X_1 + X_2)^2 \, p(X_1) p(X_2) \\
&= \sum_{X_1} \sum_{X_2} \left(X_1^2 + 2 X_1 X_2 + X_2^2 \right) p(X_1) p(X_2) \\
&= \sum_{X_1} X_1^2 p(X_1) + 2 \left(\sum_{X_1} X_1 p(X_1) \right) \left(\sum_{X_2} X_2 p(X_2) \right) + \sum_{X_2} X_2^2 p(X_2) \\
&= 1 \times 1 + 2 \times 0 \times 0 + 1 \times 1 = 2 \\
\Rightarrow D &= \sqrt{2}.
\end{aligned} \tag{3.53}$$

So, after two steps, Moe is on average a distance of $\sqrt{2}$ away from where he started.

Note that the average value of $(X_1 + X_2)^2$ in this case is the same as the average value of $(X_1 + X_2 - 2\mu)^2$, since $\mu = 0$. A longer but more suggestive notation is to write that the standard deviation or uncertainty of the sum of X_1 and X_2 is

$$\Delta(X_1 + X_2) = \sqrt{2}. \tag{3.54}$$

This is an example of the general rule, first described in Equation (2.13), that the uncertainty in the sum of two quantities $x \pm \Delta x$ and $y \pm \Delta y$ is $\sqrt{\Delta x^2 + \Delta y^2}$. In this case, the uncertainty of each step is ± 1, so the uncertainty in Moe's position after two steps is $\sqrt{1+1} = \sqrt{2}$.

3.5.2 Mean and standard deviation for averages

What does this have to do with averages?

If you measure two values in the lab and find the sample mean, you have measured $\bar{x} = (x_1 + x_2)/2$. The conventional mathematical model for what you have done is to calculate properties of $\bar{X} = (X_1 + X_2)/2$. This is quite sneaky. You measured just two numbers and averaged them. Statisticians replace your simple sum with the sum of two random variables that can take all possible values in the sample space and ask you to use the random variables to interpret your experimental results. One way to think about it is that if millions of students were all to be carrying out your experiment in essentially identical ways, their results would explore all the possible values predicted by the random variables.

Suppose X is a random variable with mean μ and standard deviation σ. Since each measurement has true mean μ, you might expect that averaging two of them together will still produce a true mean μ, and this is correct. The reason that Moe the sailor goes an average distance of zero, while the experimental measurement gives an average value of μ is that Moe has no overall velocity in any direction. Everything he does is random. If he were staggering around on a barge that was moving one meter every time he took a step, then on average he would move one meter per step, and his motion would have a nonzero mean value like the mean value in Equation (3.55).

In a formal derivation to show that $\bar{X} = \mu$, the steps are just like those in Equation (3.52) (so to speak)

$$\bar{X} = \sum_{X_1} \sum_{X_2} \tfrac{1}{2}(X_1 + X_2)p(X_1)p(X_2)$$

$$= \tfrac{1}{2}\left[\sum_{X_1} X_1 p(X_1)\right]\left[\sum_{X_2} p(X_2)\right] + \tfrac{1}{2}\left[\sum_{X_1} p(X_1)\right]\left[\sum_{X_2} X_2 p(X_2)\right]$$

$$= \tfrac{1}{2} \times \mu \times 1 + \tfrac{1}{2} \times 1 \times \mu = \mu, \tag{3.55}$$

since by definition $\mu = \sum_X X p(X)$ and $1 = \sum_X p(X)$. So summing two measurements together and dividing by two has the same mean value as either measurement alone.

Now, just as one can find the standard deviation for a single variable X, or for the sum of two variables $X_1 + X_2$, one can also find the standard deviation for the average of two variables $\bar{X} = (X_1 + X_2)/2$. This case is so important that a special name attaches to it. The standard deviation, or uncertainty, of a sample mean is called the standard error.

The uncertainty or standard error of the mean $\bar{X}$ is less than the uncertainty of either measurement alone. Follow steps just like those in Equation (3.53):

$$
\begin{aligned}
(\Delta\bar{X})^2 &= \sum_{X_1}\sum_{X_2}\left(\frac{X_1 + X_2}{2} - \mu\right)^2 p(X_1)p(X_2) \\
&= \tfrac{1}{4}\sum_{X_1}\sum_{X_2}(X_1 - \mu + X_2 - \mu)^2 p(X_1)p(X_2) \\
&= \tfrac{1}{4}\sum_{X_1}\sum_{X_2}\left((X_1 - \mu)^2 + 2(X_1 - \mu)(X_2 - \mu) + (X_2 - \mu)^2\right)p(X_1)p(X_2).
\end{aligned}
$$
(3.56)

The first term of Equation (3.56) is

$$
\tfrac{1}{4}\left[\sum_{X_1}(X_1 - \mu)^2 p(X_1)\right]\left[\sum_{X_2}p(X_2)\right] = \tfrac{1}{4}\sigma^2 \times 1.
$$

The second term is

$$
\tfrac{2}{4}\left[\sum_{X_1}(X_1 - \mu)p(X_1)\right]\left[\sum_{X_2}(X_2 - \mu)p(X_2)\right] = (\mu - \mu) \times (\mu - \mu) = 0.
$$

The third term is the same as the first. Summing them together gives

$$
\Delta X = \frac{\sigma}{\sqrt{2}}.
$$
(3.57)

Equation (3.57) shows that if you have any two measurements whose uncertainty is σ and you calculate the sample mean from them, your uncertainty goes down by a factor of $\sqrt{2}$. It doesn't matter what the probability distribution p happens to be.

What about experiments where you take three measurements and average them? Well, the standard deviation of the sum of two measurements is $\Delta(X_1 + X_2) = \sqrt{2}\sigma$. If you add a third measurement X_3 with standard deviation σ, then using Equation (3.59), you get the new standard deviation by taking the square of the standard deviation of the first, plus the square of the standard deviation of the second:

$$
\Delta(X_1 + X_2) + \Delta X_3 = \sqrt{\left(\sqrt{2}\sigma\right)^2 + \sigma^2} = \sqrt{3\sigma^2}.
$$
(3.58)

Thus the standard deviation or uncertainty in the average of three measurements $(X_1 + X_2 + X_3)/3$ is

Generalization: *Suppose you add together two variables X and Y whose probability is given by distributions $p_X(X)$ and $p_Y(Y)$ with two different true means μ_X and μ_Y and two different standard deviations ΔX and ΔY. What is the standard deviation of the sum?*

$$[\Delta(X+Y)]^2 = \sum_X \sum_Y (X - \mu_X + Y - \mu_Y)^2 p_X(X) p_Y(Y)$$

$$= \sum_X \sum_Y \left((X - \mu_X)^2 + 2(X - \mu_X)(Y - \mu_Y) + (Y - \mu_Y)^2 \right) p_X(X) p_Y(Y)$$

$$= \left[\sum_X (X - \mu_X)^2 p_X(X) \right] \left[\sum_Y p_Y(Y) \right]$$

$$+ 2 \left[\sum_X (X - \mu_X) p_X(X) \right] \left[\sum_Y (Y - \mu_Y) p_Y(Y) \right]$$

$$+ \left[\sum_X p_X(X) \right] \left[\sum_Y (Y - \mu_Y)^2 p_Y(Y) \right]$$

$$= \left[\Delta X^2 \right] [1] + 2 \times [\mu_X - \mu_X][\mu_Y - \mu_Y] + [1] \left[\Delta Y^2 \right] = \Delta X^2 + \Delta Y^2$$

$$\Rightarrow \Delta(X+Y) = \sqrt{\Delta X^2 + \Delta Y^2}. \tag{3.59}$$

$$\frac{\Delta(X_1 + X_2 + X_3)}{3} = \frac{1}{3}\sqrt{3\sigma^2} = \frac{\sigma}{\sqrt{3}}. \tag{3.60}$$

And it keeps going. Add a fourth measurement to the first three. Now the uncertainty in the sum is $\Delta(X_1 + X_2 + X_3) + \Delta X_4 = \sqrt{3\sigma^2 + \sigma^2} = \sqrt{4\sigma^2}$, and the uncertainty in the average is $\Delta(X_1 + X_2 + X_3 + X_4)/4 = \sqrt{4\sigma^2}/4 = \sigma/\sqrt{4}$.

Continuing in this fashion, one finds that the standard deviation of a drunken walker from his starting point after N steps is $\sigma\sqrt{N}$. The uncertainty in a sample mean produced from N measurements is $\sqrt{N}$ times smaller than the uncertainty of one measurement:

$$\frac{\Delta(X_1 + X_2 + \cdots + X_N)}{N} = \Delta \bar{X} = \frac{\sigma}{\sqrt{N}}. \tag{3.61}$$

This result is amazingly general. It is true for any value of N. It is true for any probability distribution p. The only assumption requiring care is that the different measurements $X_1, \ldots, X_N$ are truly independent so that the random differences between them cancel out. It was derived for a discrete distribution, but exactly the same arguments hold for continuous distributions. For example, Equation (3.59) can be obtained for continuous distributions through the calculation ending in Equation (3.62).

Further generalization: *Suppose you add together two independent variables X and Y with two different true means μ_X and μ_Y and two different standard deviations ΔX and ΔY, where the probability distributions are continuous. What is the standard deviation of the sum?*

$$[\Delta(X+Y)]^2 = \int \int (X+Y-\mu_X-\mu_Y)\mathcal{P}_X(X)\mathcal{P}_Y(Y)dXdY$$

$$= \int \int \left[\begin{array}{c} (X-\mu_X)^2 \\ + \quad 2(X-\mu_X)(Y-\mu_Y) \\ + \quad (Y-\mu_Y)^2 \end{array} \right] \mathcal{P}_X(X)\mathcal{P}_Y(Y)dXdY$$

$$= \int (X-\mu_X)^2\mathcal{P}_X(X)dX \times \int \mathcal{P}_Y(Y)dY$$

$$+2\int (X-\mu_X)\mathcal{P}_X(X)dX \times \int (Y-\mu_Y)\mathcal{P}_Y(Y)dY$$

$$+\int \mathcal{P}(X)dX \times \int (Y-\mu_Y)^2\mathcal{P}_Y(Y)dY$$

$$= \Delta X^2 \times 1 + 2 \times 0 \times 0 + 1 \times \Delta Y^2$$

$$\Rightarrow \Delta(X+Y) = \sqrt{\Delta X^2 + \Delta Y^2}. \qquad (3.62)$$

Main point: Suppose the standard deviation of X from the true mean is σ. Then the standard deviation of the sample mean $\bar{X} = (X_1 + X_2 + \cdots + X_N)/N$ from the true mean is exactly $\sigma/\sqrt{N}$.

There is one problem with Equation (3.61). When you are conducting an experiment, there is no way to know the population standard deviation σ for sure. However this is not a deadly problem, because you can estimate it from the sample standard deviation. Rewriting Equation (3.61) with the approximation $s \approx \sigma$, and using Δx instead of $\Delta \bar{x}$ to refer to the standard error in the mean of x gives one of the most important results in statistics:

$$\Delta x \approx \frac{s}{\sqrt{N}}. \qquad (3.63)$$

Application of standard error: Improving an experiment

Suppose you have carried out an experiment where you grow tomatoes with EZGrow fertilizer and compare them with tomatoes grown without it. You have 15 plants grown with EZGrow from each of which you randomly choose one tomato and 13 plants grown without EZGrow from which you also randomly take one.

You find that the sample mean for the size of tomatoes grown with the fertilizer is 10 cm, with sample standard deviation 2 cm, standard error 2 cm/$\sqrt{15}$ = 0.5 cm, and that for tomatoes grown without the fertilizer, the sample mean is 9 cm with sample standard deviation 2.5 cm, and standard error 2.5 cm/$\sqrt{13}$ = 0.7 cm. The separation between the sample means of 1 cm is less than the sum of the two standard errors (1.2 cm), and the experiment is not conclusive. You decide to construct a new version of the experiment that will enable you to distinguish the effects of the fertilizer even if the real difference between the mean sizes of the tomatoes is only half a centimeter. Let N be the number of tomatoes in each treatment group in the new experiment. Then

$$\frac{2\text{ cm}}{\sqrt{N}} + \frac{2.5\text{ cm}}{\sqrt{N}} < 0.5\text{ cm}$$

$$\Rightarrow \frac{4.5\text{ cm}}{0.5\text{ cm}} < \sqrt{N} \Rightarrow N > 81.$$

The new level of accuracy requires around 100 tomato plants in each treatment group, and since around five tomatoes grow on each plant, you buy $20 + 20 = 40$ plants.

3.6 Central Limit Theorem

It is remarkable that from the sample standard deviation, you can estimate how the standard error will decrease as you take more and more measurements. But there is an even more surprising result. When you have combined N measurements $x_1, \ldots, x_N$ together in order to obtain the sample mean $\bar{x}$, the value you obtain will almost always differ somewhat from the true mean μ. If measurements X_i are sampled from a distribution $\mathcal{P}$ with mean μ and standard deviation σ, then the sum

$$\bar{X} = \frac{\sum_{i=1}^{N} X_i}{N} \tag{3.64}$$

is a new random variable with mean μ and standard deviation $\sigma/\sqrt{N}$. Now here is the amazing thing. Not only can one say that the uncertainty in $\bar{X}$ is $\sigma/\sqrt{N}$, but as N becomes large, the probability of taking a measurement and finding some particular value of $\bar{X}$ is precisely

$$\mathcal{P}(\bar{X}) = \frac{e^{-(\bar{X}-\mu)^2/(2\sigma^2/N)}}{\sqrt{2\pi\sigma^2/N}} = \frac{e^{-(\bar{X}-\mu)^2/(2\Delta X^2)}}{\sqrt{2\pi\Delta X^2}}. \tag{3.65}$$

This is the Central Limit Theorem. The particular function that shows up in it is called the Gaussian, normal, or bell curve as first mentioned in Section 3.3.3. In application to the sample mean, the Gaussian distribution has width $\Delta x = \sigma/\sqrt{N}$ rather than width σ. The typical distance that $\bar{x}$ can vary away from μ before the probability drops to very small values is Δx. The function $\mathcal{P}(x)$ is largest when $\bar{x}$ equals μ and falls off when $\bar{x}$ is either larger or smaller than this value. So although the true mean μ is not the only value of the sample mean $\bar{x}$ one can measure, it is the single most likely value.

What does it mean? Like all continuous probability distributions, $\mathcal{P}(\bar{X})d\bar{X}$ gives the likelihood that $\bar{X}$ will lie in the range $\bar{X}$ and $\bar{X} + d\bar{X}$. Thus Equation (3.65) gives the likelihood of measuring every possible value for the sample mean $\bar{X}$. One way you could interpret this function is by thinking of a classroom where

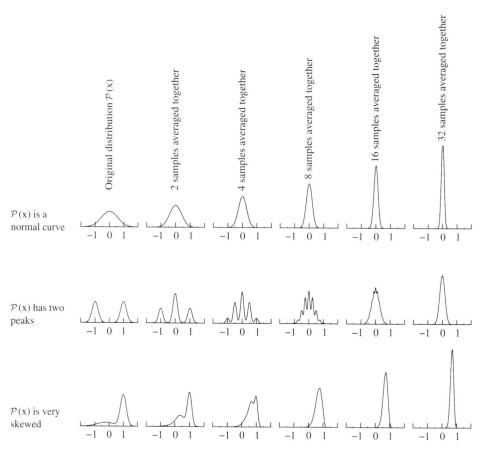

Figure 3.6 No matter what distribution you start with, if you take samples from it and average them together, the probability distribution of the sample means turns to a normal distribution that becomes narrower and higher as the number of terms in the sample mean increases. This figure illustrates the idea by showing on the left three different probability distributions. Moving to the right, the figure shows the probability distribution of the sample mean after one has averaged together 2, 4, 8, 16, and 32 samples. By the time the sample mean is built from 32 measurements, the distribution in each case is very close to a normal curve, centered on the mean of the curve farthest to the left, and with a standard deviation that is $\sqrt{32}$ times smaller than the standard deviation of the curve farthest to the left.

each student has been asked to measure something N times and obtain a sample mean $\bar{X}$. The students' answers for the sample means will all be different, just as their original data are all different. If the students create a histogram of their sample means, then as the number of students contributing becomes very large, the result will be proportional to the probability distribution function $\mathcal{P}(X)$.

When you carry out an experiment to measure some quantity $\bar{x}$, you measure the sample mean just once, yet the Central Limit Theorem lets you reach conclusions about what would have happened if you had repeated your experiment from start to finish many times.

The result is particularly extraordinary because the Gaussian probability of measuring different values of $\bar{X}$ is correct no matter what the original probability $\mathcal{P}$ of measuring the variables X_i may have been. Figure 3.6 shows three different probability distributions, and how the probability distribution for $\bar{X}$ looks after 2, 4, 8, 16, and 32 samples have been combined into a sample mean. No matter the shape of the curve on the left, the curves on the right are all normal curves. In the examples shown in the figure, a sample mean with 32 data points is enough to ensure that the result is normally distributed. In fact, for these examples, the normal curve starts to become a decent approximation somewhere between a sample size of 4 and 8. In practice, one often demands a sample size of at least five before beginning to make use of normal statistics.

The proof of the Central Limit Theorem is contained in books on probability and statistics, such as Feller (1968). In this class, it is more important to know how to use the results of the Central Limit Theorem than to know how to prove it.

Application of the normal distribution to natural data sets
Natural phenomena like people's heights, the sizes of storms, or the strength of earthquakes are made from sums of many smaller events, so they should be normally distributed. Right? Well, actually no. The Central Limit Theorem says that sample means are normally distributed when the samples are large enough. It makes no statement about the distributions of natural phenomena. In general naturally occurring quantities are **not** normally distributed. For example, Figure 3.7 shows the probability that Southern California earthquakes measured over a three-year period have a certain size on the Richter scale, compared with the best-fit normal distribution. Plotted on a linear scale, the normal distribution and the histogram of earthquakes look similar; they both have a peak around magnitude 2. But on a logarithmic scale, the failure of the normal distribution becomes more apparent. Earthquakes of magnitude 5 are 1000 times more likely to happen than the normal distribution would predict, and as the magnitude of the earthquake grows, so does the discrepancy.

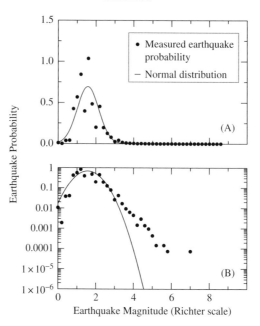

Figure 3.7 (A) Probability of Southern California earthquakes (1997–2000) as a function of their magnitude on the Richter scale plotted alongside the closest fitting normal distribution. (B) Same data as in (A) but now the vertical scale is logarithmic to reveal the probabilities of rare events.

3.6.1 Uses of the normal distribution

Once you know that the probability of measuring a sample mean $\bar{X}$ is given by a normal distribution, you can use this knowledge to test hypotheses, and find the precise likelihood your results came about by chance. To simplify calculations, notice that the argument of the exponential in the normal distribution is

$$Z \equiv \frac{\bar{X} - \mu}{\Delta X}, \text{ or from data, } Z = \frac{\bar{x} - \mu}{\Delta x}. \tag{3.66}$$

You can interpret Z as the difference between your sample mean and the true mean, measured in units of the standard error. When Z is much larger than one, your sample mean differs much more from the true mean than is likely by chance alone. When Z is much less than one, chance can easily explain the difference. Written in terms of Z, the normal distribution is

$$\mathcal{P}(Z) = \frac{1}{\sqrt{2\pi}} e^{-Z^2/2} \tag{3.67}$$

and a plot of this function appears in Figure 3.8.

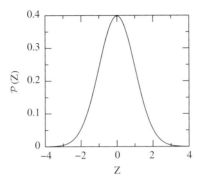

Figure 3.8 The normal distribution, plotted as a function of $Z = (\bar{x} - \mu)/\Delta x$.

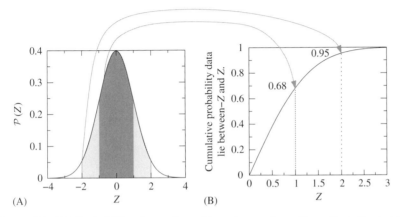

Figure 3.9 (A) Probability that $-1 < Z < 1$ or $-2 < Z < 2$, indicated by shading in the relevant parts of the normal distribution. (B) Graph of the fraction of the normal distribution contained in the interval $[-Z, Z]$ as a function of Z.

Since the normal distribution $\mathcal{P}(Z)$ is a probability density distribution, areas under the curve give probabilities. Referring to Figure 3.9, one can ask what fraction of the total probability lies in the interval $-1 < Z < 1$. The answer, obtained by integrating the normal curve from -1 to 1, is 0.68. If the interval broadens outward so that $-2 < Z < 2$, the fraction of the probability rises to 0.95. What this means is that if your data are in fact being sampled from a distribution of mean μ and standard deviation σ, and if you compute $(\bar{X} - \mu)/\Delta X = Z$, then 95% of the time Z will lie between -2 and 2, and 5% of the time by chance alone it will lie outside of this interval. An equivalent statement, following from the definition of Z, is that 95% of the time $\bar{X}$ will lie between $\mu - \Delta X$ and $\mu + \Delta X$, and 5% it will lie outside this interval. Such integrated areas under the normal distribution are very significant, so they are recorded in Tables 3.4 and 3.5. The first table gives areas under the normal distribution that lie between $-Z$ and Z. The second table gives the same information in two different ways. One column records the fraction of the

Table 3.4 Area under a normal distribution between $-Z$ and Z as a function of Z.

Z	Probability that your measured average lies between $-Z$ and Z	Z	Probability that your measured average lies between $-Z$ and Z
0.0	0.000	1.6	0.890
0.1	0.080	1.7	0.911
0.2	0.159	1.8	0.928
0.3	0.236	1.9	0.943
0.4	0.311	**1.96**	**0.950**
0.5	0.383	2.0	0.954
0.6	0.451	2.1	0.964
0.7	0.516	2.2	0.972
0.8	0.576	2.3	0.979
0.9	0.632	2.4	0.984
1.0	0.683	2.5	0.988
1.1	0.729	2.6	0.991
1.2	0.770	2.7	0.993
1.3	0.806	2.8	0.995
1.4	0.838	2.9	0.996
1.5	0.866	3.0	0.997

Table 3.5 Areas under the normal curve outside regions bordered by $\pm Z$.

Z	p-value for two-tailed test	p-value for one-tailed test	Z	p-value for two-tailed test	p-value for one-tailed test
0.0	1.000	0.500	1.6	0.110	0.055
0.1	0.920	0.460	1.7	0.089	0.045
0.2	0.841	0.421	1.8	0.072	0.036
0.3	0.764	0.382	1.9	0.057	0.029
0.4	0.689	0.345	**1.96**	**0.050**	0.025
0.5	0.617	0.309	2.0	0.046	0.023
0.6	0.549	0.274	2.1	0.036	0.018
0.7	0.484	0.242	2.2	0.028	0.014
0.8	0.424	0.212	2.3	0.021	0.011
0.9	0.368	0.184	2.4	0.016	0.008
1.0	0.317	0.159	2.5	0.012	0.006
1.1	0.271	0.136	2.6	0.009	0.005
1.2	0.230	0.115	2.7	0.007	0.003
1.3	0.194	0.097	2.8	0.005	0.003
1.4	0.162	0.081	2.9	0.004	0.002
1.5	0.134	0.067	3.0	0.003	0.001

normal distribution that does **not** lie between $-Z$ and Z. Because this fraction of the normal curve fills out two tails going to the left and to the right, the data in this column are called probabilities or p-values for a *two-tailed test*. The next column records the fraction of the normal distribution that is **greater** than Z. These data are called probabilities or p-values for a one-tailed test. If Z is negative, then the p-values for the two-tailed tests are given by $p(Z) = p(-Z)$, while the p-value for the one-sided test is given by $p(Z) = 1 - p(-Z)$. That is, for the two-sided test, ignore the sign of Z, while for the one-sided test, subtract p from one if Z is negative. There is another one-sided test that could be tabulated instead which would give the area under the normal curve that is *less* than Z. The p-values for this test are obtained by computing $1 - p(Z)$, using the values for the one-tailed test.

3.6.2 *Z test and interpreting* p

How can areas under curves be used in practice? Suppose you are gathering data and want to test the null hypothesis that the data come from a distribution with mean μ and standard deviation σ. You take N samples, compute the sample mean $\bar{x}$, the standard error $\sigma/\sqrt{N} = \Delta x$, and finally $Z = (\bar{x} - \mu)/\Delta x$. There is a conventional rule that you can reject the null hypothesis if the value you found could have arisen by chance only 5% of the time. The value of Z so that 95% of the normal curve lies between $-Z$ and Z is 1.96. This means that the area under the normal curve where Z is less than -1.96 and greater than 1.96 is 5%. Looking at Figure 3.9, this means the area under the two wings to the left and right, so using a criterion of this sort is called a *two-tailed test*. You look at the value of Z you have obtained and compare it with 1.96. If your Z value is less than -1.96 or greater than 1.96, you say

"If the null hypothesis is true, if chance alone is operating, and if I were to repeat the entire experiment many times, Z values this different from zero would appear only 5% of the time. Therefore I reject the null hypothesis."

Remember that although the *idea* of repeating the experiment many times is part of interpreting p, in fact you perform the complete experiment only once. This analysis of data is called a Z test, and the probability p can be found in the column labeled "p-value for two-tailed test" in Table 3.5.

3.6.3 *Large Z, small p*

If your measurements were to deviate more and more from the true mean, Z would get larger and larger. At the same time, as the difference between the sample mean and true mean gets larger and larger, p gets smaller and smaller. Large Z means "reject the null hypothesis." Small p means "reject the null hypothesis." The reason to introduce both Z and p is that Z is easy for you to compute directly from your data. However it is not easy to say exactly what it means. Once you have Z, you can look up p in a table or find it from a computer, and p has a very precise interpretation.

Imprecise and incorrect interpretations of $p < 0.05$

- "There is less than a 5% chance my experiment didn't work."
- "Only 5% of the time my results would be wrong."
- "I'm 95% confident my results are right."
- "My experiment works 95% of the time."

Imprecise and incorrect interpretations of $p > 0.05$

- "My results are wrong more than 5% of the time."
- "More than 5% of people would say my results are wrong."
- "I proved that my null hypothesis is right."
- "My experiment is not significant and I have no results."

Better interpretation of $p < 0.05$

- "I reject the null hypothesis (at the 5% level) because chance alone could produce results like mine less than 5% of the time."

Better interpretation of $p > 0.05$

- "I cannot reject the null hypothesis because chance alone could explain results like mine more than 5% of the time."

Example: Z test for 100 flips of a coin

Return to the data obtained by flipping a coin 100 times in Section 2.1.6. If the coin is fair, the true mean of the coin flips is $\mu = 0.5$, and the standard deviation $\sigma = 0.5$. The goal of the experiment is to test the null hypothesis that the coin is indeed fair. Since the true mean and standard deviation are known, the Z test is appropriate. After 100 flips, $N = 100$, the sample mean is $\bar{x} = 0.46$, and the standard error is $\Delta x = 0.05$. To form Z, compute

$$Z = \frac{0.46 - 0.5}{0.05} = 0.8.$$

Examining Table 3.5, the fraction of the area under the normal curve where Z is greater than 0.8 in absolute value is $p = 0.42$. So if the coin is fair, 42% of the time Z would differ from zero as much as it does here or more. Since this degree of deviation is so likely purely by chance there is no reason to reject the null hypothesis. So far as you can tell, the coin is fair.

In most cases you should employ a two-tailed test to evaluate the probability your observations have arisen by chance. There is only one instance when the use of a one-tailed test is justified. This case occurs when when you have certain knowledge that the value you are measuring must be greater (or less) than the value with which you are comparing it. So, even if you measured a sample mean that was more than two standard errors below the expected value employed by your null hypothesis, you would confidently assert that your data arose due to chance alone. Having this sort of confidence is rare. It is not justifiable to use a one-sided test just because you have the opinion, expectation, or desire that one number will be bigger than the other. It is justifiable when, because of extra information obtained from sources other than your experiment, you can rule out some experimental outcomes and therefore need less data to arrive at a confident conclusion than you would otherwise.

3.6.4 Two-tailed and one-tailed tests

Example: *A justifiable one-sided test*
Suppose you are dropping a ball and measuring its velocity when it hits the ground. You have previously measured the velocity v_{10} of this ball as it drops from a height of 10 m and repeated the experiment so often and with such care that the uncertainty is less than ±0.01 m/s. Now you repeat the experiment from a height of 9.5 m but have time for many fewer trials, and on some occasions find a velocity at the ground that is larger than the value you have obtained in dropping the ball from 10 m. However, you are absolutely certain that the ball can only speed up as it falls and that the actual speed when the ball drops from 9.5 m must be less than the actual speed v_{10} when it drops from 10 m. There is the possibility that after averaging together a number of values, the result would still be greater than v_{10} but you would be completely confident in rejecting this result as statistical error, and therefore can be justified in using a one-tailed test.

3.6.5 Confidence intervals

There is a different way that one can interpret the mathematics of the normal distribution. Instead of assuming that the true mean μ of the underlying probability

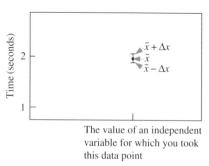

Figure 3.10 Error bars rise a distance Δx above and below the sample mean $\bar{x}$. They can be interpreted in two ways. One interpretation is that if the true mean μ is equal to $\bar{x}$, then 68% of the time people measuring $\bar{x}$ will find it in this range. The other interpretation is that the true mean μ has a 68% chance to lie in this range.

distribution is known and using that knowledge to calculate the probability of obtaining a sample mean $\bar{X}$, turn things around. Discuss the probability that the real underlying distribution has true mean μ, for different values of μ.

From this point of view, suppose you have conducted an experiment, and find $\bar{x}$ and Δx. You know that $\mu = \bar{x} - \Delta x \times Z$. You know from Table 3.5 that 95% of the time, Z lies between -1.96 and 1.96. So 95% of the time, μ lies between $\bar{x} - 1.96\Delta x$ and $\bar{x} + 1.96\Delta x$. This interval is called the 95% *confidence interval*. You are confident that 95% of the time, the true mean would lie in this range. The bigger the confidence interval, the larger the range in which the true mean might fall. The smaller the confidence interval, the more precisely it is specified. So a large confidence interval means that you have little confidence that you know the true mean precisely, while a small confidence interval means that you have high confidence you know the true mean precisely.

Error bars provide a graphical representation of confidence intervals. The most common convention is to draw error bars of height Δx, as in Figure 3.10. An error bar of height Δx corresponds to a Z value of 1. From Table 3.4, 68% of the time μ should lie within the interval between $\bar{x} - \Delta x$ and $\bar{x} + \Delta x$. So, if your sample mean $\bar{x}$ contains enough points N that normal statistics apply, standard error bars show an interval that contains the true mean 68% of the time.

3.6.6 Type I errors

Normal statistics makes it possible to make quantitative estimates of the probability of Type I errors (Section 2.1.7). Suppose you have carried out an experiment to test some hypothesis and find a value of $Z = 2.1$. According to Table 3.5, only 3.6% of the time could chance alone explain your results. Since $p < 5\%$, you are conventionally allowed to reject the null hypothesis. This means that most scientists will allow you to publish the statement that chance alone does not explain your observations. They will agree you found a real effect.

Application of confidence intervals: *Margin of error in a survey*
You conduct a survey of 100 people, 54% of whom say they support building a new toll road round the city, and 46% of whom say they oppose it. You want to estimate the margin of error to see if the survey should represent opinions in the whole city, or whether the results could be misleading because of the small sample. View the survey as an estimate of the fraction $f = 0.54$ of the population that favors the toll road. Using Equation (3.34), you can estimate the standard deviation as

$$\sigma \approx \sqrt{0.54 \times 0.46} = 0.50;$$

this is an estimate because you do not know f exactly. As a rule of thumb, using $\sigma \approx 0.5$ for the standard deviation for responses to questions with two responses is a good starting point. The standard error is

$$\Delta f = \frac{0.5}{\sqrt{100}} = 0.05.$$

In reporting the results of surveys, the custom is to compute a 95% confidence interval, and report it as the *margin of error*. The upper limit of the 95% confidence interval is

$$f = 0.54 + 0.05 \times 1.96 \approx 0.5 + 0.1 = 50\% + 10\%$$

and the lower limit is

$$f = 0.54 - 0.05 \times 1.96 \approx 0.5 - 0.1 = 50\% - 10\%.$$

Newspapers will typically report this margin of error as "ten percentage points." Since the margin of error is larger than the difference between the fractions 0.54 and 0.46 you found for and against the road, your survey did not sample enough people to determine if a majority really favors the new road.

But please be careful. If you reject the null hypothesis, you must use judgment, because there is still a 3.6% chance that you will believe in an effect when there isn't one, and make a Type I error. If your study is trying to determine if a new teaching method will raise student scores in learning algebra, you are probably willing to use the new method knowing there is a 3% risk that scores will not go up after all. However, if you are screening for a disease that requires an operation, you should demand additional and more precise tests if there is a 3% chance you could be wrong. The requirement that $p < 5\%$ is really not a very demanding standard after all. Out of 100 papers published by perfectly good scientists claiming to have found some new scientific effect at around 95% confidence, five papers and five conclusions are likely to be wrong. Why not adopt a higher standard then? Every time you cut in half the probability that chance alone could explain your data, you must gather four times as much data. Someone has to pay for that. Through long experience, scientists have settled on $p < 0.05$ as a good balance between the cost of doing experiments and the cost of getting wrong answers.

Example: *Mammograms*

Mammograms are a common screening procedure to determine if women have breast cancer, but they are also prone to errors. 10.7% of mammograms in the U.S. give false positives (Type I error) (Jatoi *et al.* 2006), that is, they detect a breast cancer which is not really present. More detailed analysis of tissue samples prevents most unnecessary operations, but there is still psychological stress associated with the mistaken diagnosis. Around 2% of mammograms give false negatives (Type II error), and fail to detect cancers they should have found. These missed cancers result in a large proportion of the medical malpractice suits in the United States (Reintgen *et al.*, 1993).

3.6.7 *Z tests in spreadsheets*

Spreadsheets provide several ways to compute Z tests. The most reliable way is for you to compute Z yourself and then get the p-value from $= (1-2*\text{GAUSS}(Z))$. There is also a function `ZTEST` that acts directly upon your data and gives you a p-value in return. However, beware of this function because in many spreadsheets the documentation says that `ZTEST` returns the p-value for a two-tailed test, while in fact it returns the p-value from a one-tailed test.

3.7 Comparing many experimental measurements

There is a wide variety of statistical tests in addition to the Z test. The reason that the Z test is not sufficient is that it only applies to cases where one wishes to compare a single experimental sample mean $\bar{x}$ with a known value μ. There are many other sorts of comparisons that arise during experiments. However, these additional tests have the same basic form as the Z test. The structure of each test has two parts. First, you extract some number from your data that is similar to Z in that it involves different quantities you want to compare. Next, you assume that your data are described by some statistical model and calculate from the model the probability that this number could assume different values. By comparing the number you have actually obtained with its statistical likelihood, you decide whether or not your results are due to chance.

3.7.1 *t test: Comparing two measurements*

The Z test is only applicable when both the true mean and standard deviation of the normal distribution with which you wish to compare are known. You may sometimes know a mean value that you want to compare with your data, but need to estimate the standard deviation.

The t test tells you the likelihood that an average $\bar{x}$ you have measured comes from a distribution with mean μ; you no longer assume that you know the standard deviation in advance.

In very close analogy with Z, define t by calculating the sample standard deviation and then computing

$$t = \frac{\bar{x} - \mu}{s/\sqrt{N}} = \frac{\bar{x} - \mu}{\Delta x}. \tag{3.68}$$

Just like Z, t is the ratio between [how far the sample mean lies from the mean you expect] and [the standard error]. If this ratio is large, your data may be telling you that you were expecting the wrong mean. It differs from Z only because you do not know the standard deviation σ in advance, but have to estimate it from the data. In contrast to Z, the probability that your results occured by chance depends not only upon Z, but also upon the number of *degrees of freedom,* which in this case is the same quantity appearing in the denominator of the sample standard deviation:

$$\text{degrees of freedom} = N - 1. \tag{3.69}$$

Table 3.6 contains p-values for different values of t and a number of different degrees of freedom. When degrees of freedom are missing, use nearby values for an estimate. For example, when $t = 2.6$, with 50 degrees of freedom, p must lie between 0.014 and 0.011, and is around 0.013. For large d, the probability of t is the same as the probability of Z.

The probability of measuring t for different degrees of freedom is plotted in Figure 3.11. Spreadsheets can compute these p-values with the function TDIST.

Statistics

Table 3.6 *p*-values for two-tailed *t* test with varying numbers of degrees of freedom.

	Degrees of freedom $d \rightarrow$						
t	4	6	10	15	20	30	100
0.0	1.000	1.000	1.000	1.000	1.000	1.000	1.000
0.1	0.925	0.924	0.922	0.922	0.921	0.921	0.921
0.2	0.851	0.848	0.845	0.844	0.844	0.843	0.842
0.3	0.779	0.774	0.770	0.768	0.767	0.766	0.765
0.4	0.710	0.703	0.698	0.695	0.693	0.692	0.690
0.5	0.643	0.635	0.628	0.624	0.623	0.621	0.618
0.6	0.581	0.570	0.562	0.557	0.555	0.553	0.550
0.7	0.523	0.510	0.500	0.495	0.492	0.489	0.486
0.8	0.469	0.454	0.442	0.436	0.433	0.430	0.426
0.9	0.419	0.403	0.389	0.382	0.379	0.375	0.370
1.0	0.374	0.356	0.341	0.333	0.329	0.325	0.320
1.1	0.333	0.313	0.297	0.289	0.284	0.280	0.274
1.2	0.296	0.275	0.258	0.249	0.244	0.240	0.233
1.3	0.263	0.241	0.223	0.213	0.208	0.204	0.197
1.4	0.234	0.211	0.192	0.182	0.177	0.172	0.165
1.5	0.208	0.184	0.165	0.154	0.149	0.144	0.137
1.6	0.185	0.161	0.141	0.130	0.125	0.120	0.113
1.7	0.164	0.140	0.120	0.110	0.105	0.099	0.092
1.8	0.146	0.122	0.102	0.092	0.087	0.082	0.075
1.9	0.130	0.106	0.087	0.077	0.072	0.067	0.060
2.0	0.116	0.092	0.073	0.064	0.059	0.055	0.048
2.1	0.104	0.080	0.062	0.053	0.049	0.044	0.038
2.2	0.093	0.070	0.052	0.044	0.040	0.036	0.030
2.3	0.083	0.061	0.044	0.036	0.032	0.029	0.024
2.4	0.074	0.053	0.037	0.030	0.026	0.023	0.018
2.5	0.067	0.047	0.031	0.025	0.021	0.018	0.014
2.6	0.060	0.041	0.026	0.020	0.017	0.014	0.011
2.7	0.054	0.036	0.022	0.016	0.014	0.011	0.008
2.8	0.049	0.031	0.019	0.013	0.011	0.009	0.006
2.9	0.044	0.027	0.016	0.011	0.009	0.007	0.005
3.0	0.040	0.024	0.013	0.009	0.007	0.005	0.003

3.7.2 Two-sample t test

The two-sample *t* test lets you test the null hypothesis that two quantities you have measured are the same. Suppose you have made N_x measurements $x_1, \ldots, x_{N_x}$ of a first quantity, and N_y measurements $y_1, \ldots, y_{N_y}$ of a comparable but potentially different quantity. The two-sample *t* test addresses the question of how likely it is to obtain a difference between the sample means $\bar{x} - \bar{y}$ assuming that the true means of the two quantities are the same.

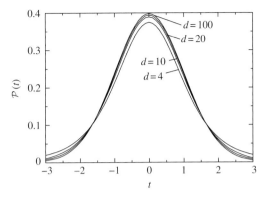

Figure 3.11 The probability of observing t is shown for $d = 4$, 10, 20, and 100 degrees of freedom. For 100 degrees of freedom, the curve is not distinguishable from the normal distribution. For small degrees of freedom, the t distribution is shorter and wider than the normal distribution. The reason is that the standard error is being estimated from the data, rather than being known in advance. When many data points happen to be close together, the standard error is estimated to be small, and t has an enhanced probability of being large. This effect goes away as the number of degrees of freedom becomes large. For large d, the probability of t is the same as the probability of Z.

Let $\bar{x}$ and $\bar{y}$ be the sample means of the N_x and N_y measurements of x and y. Let $s_{\bar{x}}$ and $s_{\bar{y}}$ be the sample standard deviations, and let $\Delta x = s_{\bar{x}}/\sqrt{N_x}$ and $\Delta y = s_{\bar{y}}/\sqrt{N_y}$ be the standard errors.

Now you have to make another choice. You have to decide whether it is possible that the true means of x and y are the same, but the standard deviations are different. If this is possible, then you should compute t from

$$t = \frac{\bar{x} - \bar{y}}{\sqrt{\Delta x^2 + \Delta y^2}}. \tag{3.70}$$

Use this quantity in Table 3.6 with the smaller of N_x and N_y for the number of degrees of freedom. This procedure provides a conservative estimate of the probability, and it puts the probability of having found an effect on the low side. The procedure is more prone to false negatives than to false positives.

If on the other hand you are confident that the standard deviations of x and y are the same, then you may define the *pooled standard error* s_D

$$s_D = \sqrt{\frac{\sum_{i=1}^{N_x}(x_i - \bar{x})^2 + \sum_{i=1}^{N_y}(y_i - \bar{y})^2}{N_x + N_y - 2}\left(\frac{1}{N_x} + \frac{1}{N_y}\right)}, \tag{3.71}$$

compute t from

$$t = \frac{\bar{x} - \bar{y}}{s_D}, \tag{3.72}$$

and employ $N_x + N_y - 2$ degrees of freedom in Table 3.6.

Spreadsheets implement both of these different t tests in the function TTEST, which acts directly upon the raw data and gives a p-value.

Example: Rolling balls

The data in Table 2.2 for balls rolling down lubricated and unlubricated channels present a case where it seems very likely that fluctuations in the data should be the same, even if the sample means are slightly different. Fluctuations are mainly due to variations in how the ball is released, and there is no reason to think they should depend upon the lubrication. Therefore, to see if the mean fall time of the ball is affected by lubrication, use Equation (3.72) with the pooled standard error. The sample mean times for the balls have already been computed in Section 2.1.5. Computing the pooled standard deviation gives

$$s_D = 0.007 \text{ sec}$$

leading to

$$t = \frac{0.324 - 0.305}{0.007} = 2.7.$$

There are 18 degrees of freedom $(10 + 10 - 2)$; from Table 3.6 on page 94, the probability p that the two sample means really come from the same distribution is between 0.016 and 0.014. Using the Excel function TTEST gives a value of 0.014. Less than 2% of the time could chance alone produce such a difference between the two means, so one can reject the null hypothesis and conclude that lubricating the channel really makes a difference.

Example: Fish in two lakes

Section 2.1.5 showed that the data in Table 2.3 can be summarized by sample mean and standard errors

$$x_{\text{Ormond}} = 4.92 \pm 0.07 \text{ mm}; \quad x_{\text{Dugout}} = 4.67 \pm 0.05 \text{ mm}. \tag{3.73}$$

It is not clear that the standard deviations of the distributions of fish in the two lakes should be the same, so to be conservative, assume they are not and use Equation (3.70). The value of t is

$$\frac{4.92 - 4.67}{\sqrt{0.07^2 + 0.05^2}} = 2.91.$$

The smallest sample of fish has 96 fish, so use this number for the degrees of freedom. From Table 3.6, the p-value is between 0.005 and 0.007, while the Excel function TTEST gives $p = 0.0056$. It is safe to reject the null hypothesis that the two fish populations have the same average size.

3.8 Data with many values of independent variable

3.8.1 χ^2 test for many means with uncertainties

The χ^2 test[3] provides a way to compare large numbers of separate measurements with a theory or with each other. It comes in two separate forms that are useful for different purposes.

For the first form of the test, suppose you have measured d different things, $\bar{y}_1, \ldots, \bar{y}_d$. Here $\bar{y}_1$ is not just one measurement. It is something you have measured many times, and for which you have found an uncertainty Δy_1, either from the standard error, or from an estimate of experimental uncertainties. Similarly $\bar{y}_2$ and all the other mean values come from repeated measurement. You cannot combine all of them together into a single average because they are different quantities.

At the same time, you have some expectation of what each measurement should give. Perhaps the expectation comes from a theory, or perhaps from previous experiments. The value you expect to have seen for $\bar{y}_1$ is μ_1, the value you expect to see for $\bar{y}_2$ is μ_2, and so on. But, as is always the case in experiments, expectations and measurements do not exactly agree. Some of the measurements $\bar{y}_i$ differ from the expected values μ_i more than others, and it is not obvious whether overall measurements and expectations agree. This is the situation calling for the first form of χ^2. Define

$$\chi^2 \equiv \sum_{i=1}^{d} \frac{(\bar{y}_i - \mu_i)^2}{(\Delta y_i)^2}. \tag{3.74}$$

Roughly speaking, if χ^2 is much larger than d, it is unlikely that chance alone can explain the deviation between $\bar{y}_i$ and μ_i, while otherwise, chance alone could well be operating. Just like the t test, a test for the probability of χ^2 requires the number of degrees of freedom in the test. When χ^2 has been computed according to Equation (3.74), the number of degrees of freedom is just the number of terms in the sum, d. The p-values needed to test hypotheses with a χ^2 test appear in Table 3.7. The entries in the table are p-values used in testing hypotheses for various numbers of degrees of freedom d. The left-most column records not χ^2, but χ^2/d. So,

---

[3] χ is a Greek letter pronounced "ki" as in "kite" in English and "chee," with the "ch" as in "aach!" in Greek. This is not the letter "x."

Table 3.7 Probability that χ^2 is larger than selected values.

χ^2/d	Degrees of freedom $d \rightarrow$										
	1	2	3	4	5	10	20	30	40	50	100
0.0	1.00	1.00	1.00	1.00	1.00	1.00	1.00	1.00	1.00	1.00	1.00
0.1	0.75	0.90	0.96	0.98	0.99	1.00	1.00	1.00	1.00	1.00	1.00
0.2	0.65	0.82	0.90	0.94	0.96	1.00	1.00	1.00	1.00	1.00	1.00
0.3	0.58	0.74	0.83	0.88	0.91	0.98	1.00	1.00	1.00	1.00	1.00
0.4	0.53	0.67	0.75	0.81	0.85	0.95	0.99	1.00	1.00	1.00	1.00
0.5	0.48	0.61	0.68	0.74	0.78	0.89	0.97	0.99	1.00	1.00	1.00
0.6	0.44	0.55	0.61	0.66	0.70	0.82	0.92	0.96	0.98	0.99	1.00
0.7	0.40	0.50	0.55	0.59	0.62	0.73	0.83	0.89	0.92	0.95	0.99
0.8	0.37	0.45	0.49	0.52	0.55	0.63	0.72	0.77	0.81	0.84	0.93
0.9	0.34	0.41	0.44	0.46	0.48	0.53	0.59	0.62	0.65	0.67	0.75
1.0	0.32	0.37	0.39	0.41	0.42	0.44	0.46	0.47	0.47	0.47	0.48
1.1	0.29	0.33	0.35	0.35	0.36	0.36	0.34	0.32	0.31	0.29	0.23
1.2	0.27	0.30	0.31	0.31	0.31	0.29	0.24	0.21	0.18	0.16	0.08
1.3	0.25	0.27	0.27	0.27	0.26	0.22	0.17	0.13	0.10	0.08	0.02
1.4	0.24	0.25	0.24	0.23	0.22	0.17	0.11	0.07	0.05	0.03	0.01
1.5	0.22	0.22	0.21	0.20	0.19	0.13	0.07	0.04	0.02	0.01	0.00
1.6	0.21	0.20	0.19	0.17	0.16	0.10	0.04	0.02	0.01	0.00	0.00
1.7	0.19	0.18	0.16	0.15	0.13	0.07	0.03	0.01	0.00	0.00	0.00
1.8	0.18	0.17	0.14	0.13	0.11	0.05	0.02	0.00	0.00	0.00	0.00
1.9	0.17	0.15	0.13	0.11	0.09	0.04	0.01	0.00	0.00	0.00	0.00
2.0	0.16	0.14	0.11	0.09	0.08	0.03	0.00	0.00	0.00	0.00	0.00
2.1	0.15	0.12	0.10	0.08	0.06	0.02	0.00	0.00	0.00	0.00	0.00
2.2	0.14	0.11	0.09	0.07	0.05	0.02	0.00	0.00	0.00	0.00	0.00
2.3	0.13	0.10	0.08	0.06	0.04	0.01	0.00	0.00	0.00	0.00	0.00
2.4	0.12	0.09	0.07	0.05	0.03	0.01	0.00	0.00	0.00	0.00	0.00
2.5	0.11	0.08	0.06	0.04	0.03	0.01	0.00	0.00	0.00	0.00	0.00
2.6	0.11	0.07	0.05	0.03	0.02	0.00	0.00	0.00	0.00	0.00	0.00
2.7	0.10	0.07	0.04	0.03	0.02	0.00	0.00	0.00	0.00	0.00	0.00
2.8	0.09	0.06	0.04	0.02	0.02	0.00	0.00	0.00	0.00	0.00	0.00
2.9	0.09	0.06	0.03	0.02	0.01	0.00	0.00	0.00	0.00	0.00	0.00
3.0	0.08	0.05	0.03	0.02	0.01	0.00	0.00	0.00	0.00	0.00	0.00

for five degrees of freedom, to find the probability that χ^2 could be larger than 6, compute $6/5 = 1.2$, and read off $p = 0.31$. Graphs of the probability of χ^2 appear in Figure 3.12.

Spreadsheets have a function CHIDIST that calculates the likelihood p of obtaining a value larger than χ^2 in an experiment with d degrees of freedom.

Example: Bottle rocket heights versus theory

You launch a bottle rocket with seven different volumes of water and measure the height to which it rises. You have repeated each measurement five times and

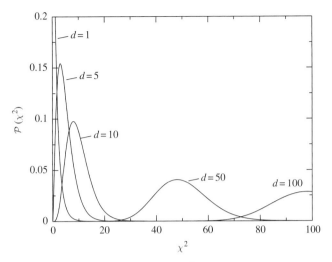

Figure 3.12 The probability of observing χ^2 is shown for various numbers of degrees of freedom d. The p-values in Table 3.7 are obtained by choosing a value of χ^2 and then computing the total area under the curve to the right of it. The graph shows only positive values of χ^2 since from its definition χ^2 must always be positive. The probability of observing a value of χ^2 peaks near the number of degrees of freedom d.

Table 3.8 Data of height versus volume of water for a bottle rocket.

Volume V (ml)	Height $\bar{y}$ (m)	Standard error Δy (m)	Expected height y (m) Equation (3.75)
100	23.37	3.93	18.48
200	32.19	3.2	33.10
300	35.08	4.64	43.88
400	57.6	7.36	50.80
500	53.04	8.76	53.88
600	56.06	11.47	53.10
700	46.56	6.27	48.48

obtained both a sample mean height y and a standard error Δy for each volume of water V, as listed in Table 3.8.

Looking at your data you decide that a parabola may provide a good fit, and you obtain the fitting function

$$y = 0.204V \text{ m/ml} - 1.925 \times 10^{-4} V^2 \text{m/ml}^2, \tag{3.75}$$

which you plot together with your data (Figure 3.13).

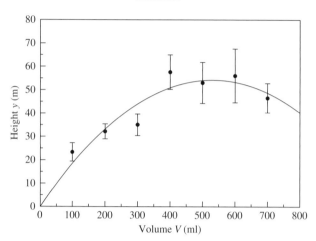

Figure 3.13 Data of height versus volume of water for bottle rocket, plotting measured values in comparison with theoretical fit.

The theoretical curve passes through some but not all of the error bars. This is not necessarily a problem. Since error bars describe a 68% confidence interval, they should bracket the true mean only around 2/3 of the time. Here, the theoretical curve lies with five of the seven error bars, so if anything the scale of the error may for some reason be overestimated.

From Equation (3.74),

$$\chi^2 = \left(\frac{23.37 - 18.48}{3.93}\right)^2 + \left(\frac{32.19 - 33.10}{3.2}\right)^2 + \left(\frac{35.08 - 43.88}{4.64}\right)^2$$
$$+ \left(\frac{57.6 - 50.80}{7.36}\right)^2 + \left(\frac{53.04 - 53.88}{8.76}\right)^2 + \left(\frac{56.06 - 53.10}{11.47}\right)^2$$
$$+ \left(\frac{46.56 - 48.48}{6.27}\right)^2 = 6.25.$$

There are seven degrees of freedom. From Table 3.7, the probability lies between 0.48 and 0.53. The Excel function `CHIDIST(6.25,7)` gives a p-value of 0.51. Thus around 50% of the time, chance alone could produce a difference between measurements and expectations where the value of χ^2 is this large or larger. There is no reason to reject the null hypothesis that the functional fit and measurements agree. In short, the fit is good.

3.8.2 χ^2 *for categories*

Under common but very special circumstances, there is a second way to compute χ^2. Assignment 3.6 sketches how to derive it from Equation (3.74). This second form is applicable whenever a data set takes the form of integers that are divided between a number of separate categories. It is particularly useful in analyzing surveys.

To use χ^2 in this way, your data must be in the form of m integers O_1, $O_2, \ldots, O_m$ which describe the number of entries you observed in categories $1, \ldots, m$. In addition, you need to come up with expectations that E_1, $E_2, \ldots, E_m$ data points should theoretically have been observed in each category. In this case,

$$\chi^2 \equiv \sum_{i=1}^{m} \frac{(O_i - E_i)^2}{E_i}. \tag{3.76}$$

There is a very specific way that Equation (3.76) can be employed to decide whether two or more groups of individuals are significantly different from one another. In this process, the expectations E_i are deduced from the observed data, and as a consequence the number of degrees of freedom used in the χ^2 test is greatly reduced. The process is best explained by example.

Three warnings about the validity of Equation (3.76)

1. If your observations are floating point numbers, or if you have converted your data from integers into percentages, then Equation (3.76) is invalid. This is not a minor quibble. The equation gives **complete nonsense.** Either you must ensure that your data have the form of integers divided between categories, or you must return to the more general Equation (3.74).

2. The mathematics lurking behind the χ^2 test is of questionable validity unless **all the observations O_i are five or more**. Sometimes one can respond to this problem by grouping categories together. However, it is quite common to conduct surveys whose results violate this guideline. If this happens you should make use of the Java applet `Survey.nlogo` on the website for this text that correctly computes probabilities even when there are fewer than five observations in some categories.

3. To use Equation (3.76) to describe data such as those in Table 3.9 through the process involving Table 3.10, every individual must contribute precisely one response to each row. The number of responses in each row must add up to the number of individuals that has been polled. If this is not the case, the results of computing χ^2 give another helping of **complete nonsense.** Just because data from a survey are assembled in matrix form does not mean the mathematics of this section is appropriate. For example, if one were to ask students to rate every course with a point score between 1 and 5, and replace Table 3.9 with a new table that summed up points for each course, the math that follows would be wrong.

Table 3.9 Results of a survey in which math and engineering majors are asked to choose a favorite math course from a list of five.

Group	Number asked	Calc 1	Calc 2	Diff EQ	Lin. Alg.
Engineering majors	300	30	45	75	150
Math majors	500	75	75	125	225

Table 3.10 Probabilities of preferring math classes assuming the null hypothesis that there is really no difference between math and engineering majors.

	Calc 1	Calc 2	Diff EQ	Lin. Alg.
Total number	105	120	200	375
Fraction	0.13	0.15	0.25	0.47

Example: Favorite math course

You have conducted a survey where you ask 300 engineering students and 500 math majors to choose a favorite math course from among Calc 1, Calc 2, Differential Equations, and Linear Algebra. You only keep survey data for students who have had all these courses.

The null hypothesis is that the fraction of people who liked a particular math course the most does not depend upon whether they are engineering or mathematics majors. You would like to test this null hypothesis and see whether in fact engineers and math majors have different favorite courses. The data in Table 3.9 provide the observations O_i, but where can the expectations E_i come from? They come from thinking carefully about the null hypothesis. If the null hypothesis is true, then it really does not matter whether someone is a math or engineering major. There is a distribution of preferences for math courses that is independent of this distinction. This distribution of preferences paying no attention to major can be obtained by summing up the total number of people who chose each math course, and then finding the fraction of the whole in each category. This process is carried out in Table 3.10.

So, assuming that math and engineering majors are really the same, out of 300 engineering majors, one would expect 13% or 39.38 to prefer Calc 1, and 15% or 45 to prefer Calc 2, and so on. These expectations are tabulated in Table 3.11, and are obtained by multiplying the fractions one would expect based on the null hypothesis in Table 3.10 by the number of people in each category. There is no problem if the expectations are not integers, although the observations must be.

Forming the sum by using Equation (3.76) for the eight observations and the expectations in Table 3.11 produces a value of $\chi^2 = 4.57$. The number of degrees

Table 3.11 Observations O_i are the upper eight entries and expectations E_i are the lower eight entries.

Group	Calc 1	Calc 2	Diff EQ	Lin. Alg.
Observed in engineering	30	45	75	150
Observed in math	75	75	125	225
Expected for engineering	39.38	45	75	140.63
Expected for math	65.63	75	125	234.38

of freedom to use in cases like this (see Section 3.8.3) is

$$d = (\text{Number of columns} - 1) \times (\text{Number of rows} - 1) \qquad (3.77)$$

where the columns and rows refer to the table of observations. In this case there are four columns and two rows, so there are three degrees of freedom. From Table 3.7, reading off the value of $\chi^2/3 = 1.52$ for three degrees of freedom gives a probability p between 0.19 and 0.21. The spreadsheet function CHIDIST(4.57,3) gives $p = 0.21$. Thus 21% of the time chance alone should be expected to produce a difference between math and engineering students as large as the one that appears in the survey, assuming that they are really the same. There are no grounds to reject the null hypothesis, and no reason to think that math and engineering majors feel differently about their favorite math course.

The spreadsheet function CHITEST acts directly on the data such as those in Table 3.11 and produces a p value.

3.8.3 Degrees of freedom

One of the mysterious aspects of statistics lies in the number of degrees of freedom d used in some of the statistical tests. They enter in t tests and χ^2 tests, and show up as the unexpected factor of $N - 1$ in the denominator of the sample standard deviation, Equation (3.7). In general, the number of degrees of freedom in a statistical calculation is

$$d = \begin{bmatrix} \text{Number of indepen-} \\ \text{dent data points} \end{bmatrix} \text{minus} \begin{bmatrix} \text{Number of independent param-} \\ \text{eters calculated from the data} \end{bmatrix}. \qquad (3.78)$$

In the case of the sample standard deviation, there are N data points, but one parameter in the formula, namely the sample mean $\bar{x}$, is computed from the data. Therefore the number of degrees of freedom is $N - 1$. For the two-sample t test in Equation (3.72), two parameters $\bar{x}$ and $\bar{y}$ are computed from the data, so the number of degrees of freedom is $N_x + N_y - 2$. When the number of degrees of freedom is small, t must be larger to indicate statistical significance than when the number of degrees of freedom is large. For use with Equation (3.70), the number of degrees of freedom is deliberately chosen smaller than is probably correct in order to make the test very conservative, and unlikely to lead to false positives.

A final example of this sort of logic explains the number of degrees of freedom used for the χ^2 test in Equation (3.77). In order to obtain the matrix of expected values, one computes a large number of parameters. At first it seems that there are as many parameters computed as data points, so the number of degrees of freedom would be zero. However not all the parameters computed are independent. All the rows in the table of expected values are proportional to one another. Begin with the parameters in the first row. They are all independent, and the number of parameters equals C, the number of columns. Go down to the next row. It is proportional to the first row, so just one more parameter is needed, namely the coefficient of proportionality. The same holds for each additional row. If there are R rows, there are $R - 1$ coefficients of proportionality. Thus the number of independent parameters in the table of expectations is $C + R - 1$. Subtracting this number of parameters from the number of data points RC gives the number of degrees of freedom

$$RC - C - R + 1 = (R - 1)(C - 1) \tag{3.79}$$

which is the same as Equation (3.77).

Unbiased estimators

The arguments of the previous section show how to compute degrees of freedom, but do not really explain why they need to be computed that way. A firm mathematical justification is provided by the idea of *unbiased estimators*. The basic point is that if you compute

$$S = \sum_{i=1}^{N}(X_i - \bar{X})^2 \tag{3.80}$$

and ask for the average value this sum S will have averaging each random variable X_i over its sample space, the answer is that the average value is not $N\sigma^2$ but $(N - 1)\sigma^2$. So dividing through by $N - 1$ gives an *unbiased estimator* of σ^2 and motivates the standard definition of the sample standard deviation.

The reason can be seen by looking carefully at the quantities in the sum in Equation (3.80). Each random variable X_i enters more than once, because not only does it occur where it is immediately visible, but it is also part of computing $\bar{X} = \sum_{j=1}^{N} X_j/N$. The sum can be rewritten as

$$S = \sum_{i=1}^{N} X_i^2 - 2 \left(\sum_{i=1}^{N} X_i \right) \left(\sum_{j=1}^{N} X_j / N \right) + N \bar{X}^2$$

$$= \sum_{i=1}^{N} X_i^2 - 2N \left(\sum_{i=1}^{N} X_i / N \right) \left(\sum_{j=1}^{N} X_j / N \right) + N \bar{X}^2$$

$$= \left(\sum_{i=1}^{N} X_i^2 \right) - N \bar{X}^2. \tag{3.81}$$

To keep the algebra a bit simpler, suppose that the true mean μ of each variable X_i is 0, and that the true standard deviation is σ. Then the average value of the first term in Equation (3.81) is $N\sigma^2$, while the average value of the second term, from Equation (3.61), is $N\sigma^2/N = \sigma^2$. Putting these together,

$$\text{average of } S = N\sigma^2 - \sigma^2 = (N - 1)\sigma^2. \tag{3.82}$$

This is the reason to define the sample standard deviation according to

$$s^2 = \frac{S}{N - 1} = \frac{1}{N - 1} \sum_{i=1}^{N} (X_i - \bar{X})^2. \tag{3.83}$$

The definition of degrees of freedom in Equation (3.78) captures the essential reason that $N - 1$ is needed, because the second term of Equation (3.81) is precisely due to the fact that the parameter $\bar{X}$ is computed from the data.

3.9 Other statistical tests

There are many other statistical tests that have not been covered here, such as ANOVA, that compares different means and assesses whether differences between them are significant. You will have to go to more complete texts on statistics to learn about them, such as Moore and McCabe (2006), Zar (1999), or Dalla (2008).

Assignments

3.1 **Inquiry – Surveying and testing**

Purpose Conduct and carry out a survey of opinion, knowledge, or learning and employ statistics to analyze the results.

Background In this inquiry, you should build upon your knowledge of statistics as you apply it to one of the most important and controversial areas of research, the investigation of people's opinions and knowledge.

Teamwork You may choose a partner for this assignment. Your partner should be someone in the same laboratory section. Together with your partner, you should settle on some question that involves other peoples' opinions or knowledge. Design a survey or assessment instrument that provides you information to answer your question.

Factors Although you have very broad freedom in what sorts of questions you address, you should include at least one potentially significant factor in your survey or test. For example, you might see whether men and women respond to something similarly or differently. You might look for variations based upon age, or on which instructor students have had.

Ethical treatment of human subjects Before beginning your survey, you will need to provide instructors with evidence that you have completed training on ethical treatment of human subjects. You must conduct this survey in accord with the guidelines that govern research on human subjects in an academic setting. In particular, you must obtain signed consent from every participant, and you must conduct the survey in such a way that the responses of all participants are anonymous, but any participant who chooses to withdraw from the study after it is completed can do so. Procedures and forms to assist in this process are on page 46.

Length The report should be 2–4 typed pages.

Report requirements Although you may design your surveys or tests in teams, please type up individual reports. The interpretation and discussion of the data should be yours, although the data may be shared.

Please include the following sections

a. Title.

b. Abstract. This paragraph should explain the purpose of your inquiry, and then summarize the main results. It should be written in present or past tense.

c. Introduction, including background information.

d. Survey. A copy of the survey or test you developed, with a discussion of why you settled on its final form.

e. Analysis. Statistical discussion of results, especially discussion of whether factors you investigated were statistically significant.

f. Conclusions.

g. Survey sheets. You must turn in the permission forms and response sheets from which you compiled your survey so that the instructors can verify that you properly followed procedures concerning ethical treatment of human subjects.

Grading This inquiry will be evaluated according to the criteria in the Inquiry Grading Rubric, Appendix D. Please note that you must acknowledge your team members, if any, and explain their contributions to the project. You must turn in your survey sheets to avoid penalties for the first item on the checklist.

3.2 **The latest diet** Dr. Dangane has a new idea for a diet, based on the following medical logic: By definition, 1 calorie is the heat it takes to raise 1 gram of water by 1 degree. If you drink 1 liter of ice water, you are warming up 1000 grams by 37 degrees, burning

37,000 calories. Since the average person eats only 2000 to 2500 calories per day, Dr. Dangane figures all his patients need to do is drink a liter of ice water, thereby burning more than 10 times their daily intake. To test this, he gets two groups of 30 volunteers who have provided signed consent forms and their current weight. Group A is asked to drink 1 extra liter of warm water at the end of each day. Group B is asked to drink 1 extra liter of ice water each day. After a month, Dr. Dangane weighs each volunteer again, and calculates the weight difference. On average, group A (warm water) weights declined by 1.24 pounds, group B declined by an average of 3.67 pounds.

Do these data justify Dr. Dangane's belief that the ice water diet will be the next big health revolution? That is, how much can one legitimately conclude given this information and no more? No calculations: just comment on the relation between data and conclusions. [For extra credit: could this diet really work? If not, what is wrong with it?]

3.3 **Mean and standard deviation** Please write down formulas for the following two values:

Sample mean: $\bar{x} =$
Sample variance: $s^2 =$

Now, answer the following questions:

a. The variance is a type of average. What is being averaged?
b. Why do you need to square a term in the equation for variance?
c. Taking the square of that term is just one solution to the problem you described in answering part (b). Can you suggest a different mathematical operation that would solve the same problem?
d. The standard deviation is defined to be the square root of the variance. Explain why it is conventional to take the square root using the following set of measurements of pH in Waller Creek to illustrate your logic:
   pH measurements: 7.34, 7.48, 7.12, 7.33, 7.28, 7.41, 7.22, 7.37, 7.30, 7.29.
e. In words, explain the difference between the standard deviation and the standard error.

3.4 **Confidence intervals for water safety** The Environmental Protection Agency regulates the amount of mercury that is allowed in drinking water, suggesting that more than 1 ppm (part per million) in water poses a health hazard that can lead to birth defects, and to brain damage in adults. A student measures mercury concentration in water samples from 20 tap water sources in Boulder to see whether Boulder water exceeds EPA standards. She obtains the following values:

1.1	1.23	1.07	0.98	0.94	1.29	1.04	0.99	1.34	1.05
1.18	1.2	1.08	1.01	0.94	1.01	0.85	0.99	1.05	1.13

Calculate the sample mean, sample standard deviation, and sample standard error for the mercury data:

$\bar{x} = $ _____ $s = $_____ $\Delta x = $_____

Based on these numbers,

a. You expect that if this experiment were repeated many times, 68.2% of the time the sample mean in ppm would lie between _____ and _____.

b. 95.5% of the time the sample mean would lie between _____ and _____.

c. And 99.7% of the time the sample mean would lie between _____ and _____.

These concentration ranges are the confidence intervals corresponding to $\pm$ 1, 2, and 3 standard errors. Provide answers accurate to two decimal places.

For the following questions, you should consider water safe when the confidence interval **includes** water considered safe by the EPA.

d. Based on the 95.5% confidence interval, is Boulder water safe?

e. Based on the 99.7% confidence interval, is Boulder water safe?

f. The student could come to a wrong conclusion in two ways:

	Water really is safe	Water really is unsafe
True status of Boulder water:		
Student decides water is okay		Type II error
Student decides water is dangerous	Type I error	

g. Which of these types of error is a bigger problem? Why?

h. Based on your answer to part (g), what confidence interval do you feel would be more appropriate? Explain your logic. [Note: consumer advocates and city budget officers might have different opinions on these last two questions!]

3.5 **A study of vitamin C** Consider the question of whether taking vitamin C supplements helps to prevent colds. Starting in July, researchers assign 2000 volunteers to take a placebo pill with no vitamin C in it, and 2000 other people to take 500 mg of vitamin C. At the end of the main cold season (autumn and early winter), they survey the volunteers as to whether or not they had a cold that year. The results look like this:

	Number of people who got a cold	Number of people who did not get a cold
Vitamin C	1200	800
No vitamin C	1250	750

a. Calculate the fraction of people with colds in each group. Fill in the table:

	Fraction of people f who got a cold
Vitamin C	
No vitamin C	

b. Consulting the examples on page 91 for hints, calculate the 95% confidence interval for the fraction of people getting colds in each group:

	Upper 95% limit	Lower 95% limit
Vitamin C		
No vitamin C		

c. Comparing the confidence intervals for the two groups, what conclusion do you reach about whether vitamin C helped prevent colds? Explain your logic.

3.6 χ^2 **test for categories** The goal of this problem is to derive and then apply a version of the χ^2 test in Section 3.8.2.

Imagine you ask a sample of N people their opinion about a topic. There are d possible answers they might give you. For instance, you ask 200 students so $N = 200$. You expect that a proportion μ_i of these people will give you answer i. For instance, you expect 25% of the students are seniors, so $i = $ "senior" and $\mu_i = 0.25$.

a. What is the number of people E_i you expect to give answer i? Write down the formula, using the symbols above.

$$E_i = \ldots \tag{3.84}$$

b. In your actual data, you observe that a *proportion* $\bar{y}_i$ actually gives you answer i. For example, 20% answer "Yes" to your first question. In general what is the observed *number* of people O_i you observe giving answer i? Write down the formula using symbols from above.

$$O_i = \ldots \tag{3.85}$$

To compute χ^2 you need to know the standard error associated with measuring $\bar{y}_i$. Statisticians have found that when N people choose result i with probability μ_i, the standard error involved in measuring $\bar{y}_i$ is

$$\Delta y_i = \sqrt{\mu_i/N}; \tag{3.86}$$

that is, the uncertainty in the fraction $\bar{y}_i$ is $\sqrt{\mu_i/N}$. This result is approximate, and depends upon N being large, but you can use it.

c. Write down the formula for χ^2 in Equation (3.74)

$$\chi^2 = \ldots \tag{3.87}$$

d. Rewrite the formula, substituting in the expression for standard error from Equation (3.86).
e. The square root and squared term in the denominator cancel out so....
f. Bring the term N up to the numerator.
g. Bring the N in the numerator inside the squared term. To do that, multiply the whole equation inside the summation symbol by $N/N = 1$.

h. Now bring the N inside the parentheses in the numerator.
   You should now have a formula that looks like

$$\chi^2 = \sum_i \frac{(N\bar{y}_i - N\mu_i)^2}{N\mu_i}. \tag{3.88}$$

i. Obtain Equation (3.76).

3.7 **Survey analysis** Imagine the following fictional survey, in which a researcher surveys local high school teachers. The researcher asks each teacher two questions:

a. Was your teacher preparation through a regular or an alternative certification program?

b. Which one of the following five areas were you least prepared for when you began teaching?

   A Subject matter knowledge;
   B Classroom management;
   C Assessment techniques;
   D Inquiry-based teaching;
   E Coping with demanding work schedule.

   The researcher is interested in knowing whether teachers trained in regular and alternative certification programs perceive their programs as having different weaknesses.

c. What is the null hypothesis for this study?

d. The following table shows the number of teachers, broken down by their preparation type, and the aspect of teaching they said they were least prepared for. This type of data is perfect for a χ^2 test in the form of Equation (3.76) because there are multiple categories of answers.

Observed numbers

Preparation area	A	B	C	D	E
Regular certification	19	20	11	16	34
Alternative certification	67	36	20	36	41

e. Following the logic in the example on page 102, fill in the following chart:

Observed numbers

Preparation area	A	B	C	D	E
All	86				

f. Convert these numbers into proportions. For instance, the total proportion of teachers answering "subject matter knowledge" is 0.2866. Fill in the rest of the following table:

Observed proportions

Preparation area	A	B	C	D	E
All	0.287				

g. Fill in the following table of expected values:

Expected proportions

Preparation area	A	B	C	D	E
Regular certification	28.66				
Alternative certification					

h. Compute χ^2.

i. Compute the probability p that χ^2 would be this large or larger by chance alone.

j. What conclusion does one draw about regularly and alternatively certified teachers?

References

G. Dalla (2008), Little handbook of statistical practice, http://www.statisticalpractice.com

Federal Election Commission (2000), http://www.fec.gov/pubrec/2000presgeresults.htm

W. Feller (1968), *An Introduction to Probability Theory and Its Applications*, vols. 1 and 2, John Wiley, New York.

I. Jatoi, K. Zhu, M. Shah, and W. Lawrence (2006), Psychological distress in U.S. women who have experienced false-positive mammograms, *Breast Cancer Research and Treatment*, **100**, 191.

H. Khamis (2008), http://www.math.wright.edu/people/harry_khamis/stature_prediction/index.htm

H. Khamis and A. F. Roche (1994), Predicting adult stature without using skeletal age: the Khamis–Roche method, *Pediatrics*, **94**, 504. Erratum: (1995) **95**, 457.

D. S. Moore and G. P. McCabe (2006), *Introduction to the Practice of Statistics*, 5th edn, Freeman, San Francisco.

C. Ogden, C. Fryer, M. Carroll, and K. Flegal (2004), Mean body weight, height, and body mass index, united states 1960–2004, http://www.cdc.gov/nchs/data/ad/ad347.pdf, *Advance Data from Vital and Health Statistics*, (347), 1–20.

D. Reintgen, C. Berman, C. Cox, *et al.* (1993), The anatomy of missed breast cancers, *Surgical Oncology*, **2**, 65.

J. Sinclair (1792), Statistical account of Scotland, v. xx, p. xiii, quoted in Encylopedia Britannica, 11th edition of 1911 in article on Sir John Sinclair, 1st baronet, available at http://en.wikipedia.org/wiki/sir_john_sinclair

J. M. Tanner, M. J. Healy, H. Goldstein, and N. Cameron (2001), *Assessment of Skeletal Maturity and Prediction of Adult Height (TW3) Method*, Elsevier, Amsterdam.

J. H. Zar (1999), *Biostatistical Analysis*, 4th edn, Pearson, London.

4

Mathematical models

4.1 Ingredients of mathematical modeling

Applied mathematics refers to the portion of mathematics most often useful in representing the physical world. Many scientists spend most of their time working on the theories of physics, chemistry, geology, astronomy, or biology; most of their time is spent with mathematics as well. Statistical analysis is a portion of that analysis, but there is much more as well. Another name for this sort of activity is *mathematical modeling.*

4.1.1 Numbers

The most useful idea in mathematical modeling is also the oldest and most basic. It is the idea of using numbers to represent the world. There are many different sorts of numbers, and it may be useful to review them before proceeding to more complicated operations.

Positive integers These represent increasing quantities of essentially identical objects. The basic idea is simple yet abstract, that one object can be essentially the same as another, while not literally being identical. Even two symbols, X, X, are different if only because they are in different places, but everyone seems to understand what it means to say that they are the same, and there are two of them.

Zero This is a number representing the absence of some quantity. It makes it possible to state that something is not present.

Negative integers These numbers seem to have been conceived in India during our Medieval period, along with the Arabic numerals (brought by Arabs from India to Europe). For a long time they were regarded as "imaginary" numbers since one cannot "really" have negative two sheep (Martinez, 2006). Today, however, the word "imaginary" is reserved for square roots of negative numbers.

Rational fractions Numbers such as 3/8 or 5/16 capture the idea of parts of a whole. They were known to ancient mathematicians. Their very name involves the idea of reason.

Irrational numbers These are numbers such as $\sqrt{2}$ or π that denote quantities but cannot be written as ratios of integers. They were also known to ancient mathematicians, who seem to have found them disturbing because they could not be written as ratios of integers, hence their name. When written in decimal format, irrational numbers need an infinite number of digits. π begins as 3.1415926535897932385..., but this is only the beginning. Irrational numbers are needed to think about quantities that change continuously, such as time, or positions in space.

Imaginary numbers All the numbers mentioned until this point are now collectively called "real numbers." Real numbers stand in opposition to "imaginary numbers," of which the primary example is $i = \sqrt{-1}$. Other imaginary numbers can be constructed through sums and products of i, for example, $(1 + i)/\sqrt{2}$, which is a square root of i. Just as people used to think you cannot "really" have -2 sheep, now they agree you cannot "really" have $2 \times \sqrt{-1}$ sheep. Describing positions in time and space does not need imaginary numbers. However the fundamental physical theory of matter, quantum mechanics, must be written using imaginary numbers. The basic equation for how particles such as electrons or protons move and interact in time needs i. This makes it hard to say that i is much less real after all than the integer 3.

4.1.2 Basic operations

The basic operations underlying mathematical modeling are addition, subtraction, multiplication, and division. Performing well-defined problems with these operations is not hard, particularly if one gets to use a calculator. What is hard is being able to select the correct operation when faced with a question about something quantitative in the world.

Here are examples where each of these basic operations would be used:

Addition

- Ellen was 13 years old at the start of her freshman year in high school. At the start of her senior year, how old is she?
- Twelve pennies sit in a jar. I put in another 15. How many are now in the jar?

Subtraction

- There are 215 Daphnia swimming in a jar on Monday morning. On Tuesday morning, seven lie dead on the bottom of the jar. How many are left alive?
- On Monday, a mold colony occupies 1.3 square centimeters. On Tuesday, it occupies 2.2 square centimeters. How much did it grow between Monday and Tuesday?

Multiplication

- There are exactly 215 Daphnia in each of eight jars. How many Daphnia do I have?
- My house has a rectangular shape 220 feet long and 80 feet wide. What is the floor space of my house?
- My car is traveling at 20 miles per hour. How far does it go after 4 hours?
- How many distinct numbers can I form with two digits, when one digit ranges from 1 to 4, and the other ranges from 5 to 9?

Division

- You have 4 ounces of plant food, and want to put the same amount in 12 pots. How much goes in each pot?
- You have 4 ounces of plant food, and want to put 0.2 ounces in pots. How many pots should you get?
- My house has 2100 square feet, and is 70 feet long. How wide is it?
- I drove 210 miles in $3\frac{1}{2}$ hours. What was my average speed?

Many problems involve operations in combination. Some examples appear in Assignment 4.1.

4.1.3 Algebra

The end result of analyzing most experiments is a collection of numbers, often represented on a graph. However it is nearly impossible to carry out the analysis working with numbers explicitly. Instead, one works with symbols such as A, g, or θ that represent numbers. There are four reasons why these symbols are so essential.

1. It is much easier to copy the symbol A without error than to copy 2.364 m^2 many times without error.
2. In many problems the actual numerical values of the symbols are not known when the problem begins; therefore, insisting on knowing the values in order to begin working the problem would lead to paralysis.
3. Using symbols rather than values means that an infinite number of problems can be solved with the same effort otherwise needed to solve just one problem.
4. Symbols begin to take on meaning after one has worked on a problem for a while, and this meaning assists in thinking about it. So m can come to signify "mass," N_f can come to signify "number of fish," and so on.

4.1.4 Setting up problems: Sketching and naming

One of the most surprising challenges you will face when analyzing an experiment arises when you sit in front of the apparatus and have to start writing down equations not provided by a lab manual or an instructor. You will need to settle on the important variables and to give them mathematical names that can later take part in equations.

For example, suppose you have tied a ball to a string, hung it from a stand, and blow air at the ball from a fan. The ball swings out at an angle because of pressure from the moving air (Figure 4.1).

Moving from the observation of such an experiment to a mathematical description of it should begin with the process of *sketching* and *naming*.

1. Draw a schematic diagram or cartoon of the experiment, as in Figure 4.2(A).
2. Indicate the essential ingredients of the experiment and name them with English words.
3. Decide which features of the experiment will need to be given quantitative values. Give each such element of the experiment a symbol, and when possible use it as a label on your sketch. Figure 4.2(B) provides an example. Sometimes an important feature of your experiment may be a function, not just a number. In this case you should label the function.
4. Formulate equations using the symbols you have defined, making it possible to represent your experiment with mathematics. For the ball on the string, the equations can come from the free-body diagram of Figure 4.2(C), which shows all the forces acting on the ball.

It would be possible in principle to employ variable names such as "Mass of Ball," and "Length of String." However names of this sort are really not useful when time comes to start carrying out algebra. The convention in science is to use

Figure 4.1 An experiment where a fan blows on a ball tied to a string.

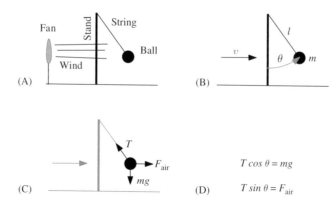

Figure 4.2 Schematic drawing of an experiment where air from a fan causes a ball on a string to swing out from the stand holding it. (A) First drawing, indicating important features of the experiment and labeling them with words. (B) Important variables in the experimental are labeled: the speed of the air is v, the angle of the ball is θ, the length of the string is l, and the mass of the ball is m. (C) Each of the forces acting on the ball is indicated by an arrow and symbols. (D) Using symbols indicated on the sketches, the balance of forces on the ball is converted into two equations, which can be solved to find the force exerted by air on the ball.

single letters to represent numerical quantities. In many cases, the letter is the first letter of the word that describes the object. So for the example in Figure 4.2 the length of the string is denoted by l, and the mass of the ball by m. Angles are often denoted by Greek symbols; in this case, the angle between the string and the stand is indicated by θ. Make sure to write out explicitly in English what every symbol means, and also make sure that you do not use the same symbol twice for two different things.

4.1.5 Functions

A function is a mapping that takes some numbers as an argument and returns other numbers as an answer. Functions can be used to draw pictures, describe trends, and gather many observations together to summarize them. There is an infinite number of possible functions, but a small number of them arise so commonly, and are so useful for building other functions, that they have special names and notation. Some are shown in Figure 4.3. Most of the laws of physics and chemistry are expressed making use of conventional functions.

Linear functions The most basic function is the straight line, $y = Ax + b$. It can be used to represent any situation where something is proportional to something else.

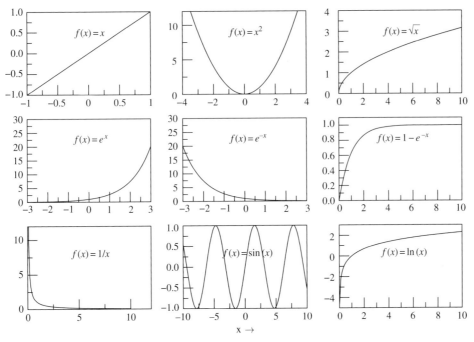

Figure 4.3 Nine common functions. Note that data corresponding to the three rightmost functions may be difficult to tell apart by eye although the functions behave very differently near zero, and at infinity.

Example: *How does the gas mileage of a car depend upon the number of passengers?* It makes sense that the mileage might decrease in proportion to the number of adults. If so, let

Concept	Symbol
Mileage	M
Mileage when only the driver is present	M_1
Number of passengers in addition to the driver	N
Adjustable mileage loss coefficient	A

Then

$$M = M_1 + AN. \tag{4.1}$$

The idea behind linear functions is very simple. Do something once, get an effect. Do it twice, get twice the effect. Linear functions can be generalized to functions of many variables,

$$y = b + \sum_{i=1}^{N} A_i x_i. \tag{4.2}$$

So, if gas mileage depends upon the weight of people in the car (x_1) and the wind speed (x_2) and the average slope of the hill the car is climbing (x_3), there might be a formula for mileage involving a linear sum of these three variables.

   If one has a relation between two variables, but no idea what the relationship should be, the first thing to try is usually a linear relationship. There is a reason to try this function first before others. Any smooth function will always look linear if one examines it over a small enough region. This statement, which is a theorem, creates the risk of a misconception, which is that whenever one provides a functional fit to experimental data, the fit should be linear. While it is reasonable to try a linear fit in the absence of anything better, it is simply not true that the only relations between quantities in nature are linear.

Integer powers Functions such as x^2 or x^7 result from multiplying an independent variable by itself a certain number of times. The function x^2 arises very naturally in one case. If you consider any smooth function near a maximum or minimum value, then if you inspect it over a small enough region, it will look like a constant times x^2. Another way to put this is that at a maximum or minimum value, the linear part of a function Ax vanishes, and what is left is quadratic. Other integer powers can occur in scientific settings, but are not particularly likely.

Negative powers Functions such as $1/x$ or $1/x^2$ arise from dividing integer powers into 1. The most likely of these functions to appear in practice is $1/x$. This function expresses the idea of inverse proportion. For example, if you have a hole in a bucket holding a certain amount of water, you might expect the time it takes for all the water to escape to be inversely proportional to the area of the hole.

Polynomials Polynomials are functions of the form $A_0 + A_1 x + A_2 x^2 + A_3 x^3 + \ldots$. A polynomial is of order n when the highest power of x that appears is x^n. There is a theorem that says that if you have n data points (x_1, y_1), $(x_2, y_2), \ldots, (x_n, y_n)$ where no two values of x_i are the same, then there is a unique polynomial of order $n-1$ that passes through the n points. This means that if you are given experimental data, you can always find a polynomial that passes through the data. So you might think that polynomials provide an all-purpose tool for modeling experiments. This is a mistake. Figure 4.4 gives an example of the problem. A polynomial fit always passes through the points one gives it, but tends to oscillate wildly between the points.

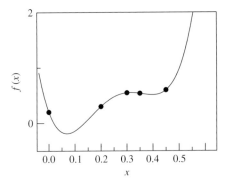

Figure 4.4 A polynomial fit will always pass through the points one gives it, but it will also usually wiggle wildly between them. This diagram shows the unique polynomial of fifth order that fits five given points. Fitting data to high order polynomials is usually a bad idea because the function wiggles wildly away from the points.

Exponentials The exponential function

$$f(t) = e^{rt} \tag{4.3}$$

arises whenever the rate at which something grows is proportional to how much there is now. The constant r gives the rate of growth. The essential behavior of the function is unchanged if one substitutes another positive constant for e. One way to see this is to note that

$$e^{rt} = p^{r't} \quad \text{where} \quad p = e^{r/r'}. \tag{4.4}$$

This function describes population growth of bacteria (or people) in the absence of limits due to food supply or disease. If something is growing exponentially, then if it doubles in time t_0, after time $2t_0$ it will have increased by four, and so on. The function grows so fast that almost nothing in the real world can increase exponentially for too long. For example, suppose that one has 10 rats, and that together they weigh 1 kg. Suppose that every year the rat population increases by 20%. Then after 312 years, the rats weigh a total of

$$1\text{kg} \times 1.2^{312} \approx 6.1 \times 10^{24}\text{kg}, \tag{4.5}$$

which is about equal to the entire weight of the Earth.

Logarithms The logarithm is the inverse of the exponential:

$$\ln[e^t] = t. \tag{4.6}$$

Logarithms can also be defined as inverses of other power functions. For example, the log base 10 obeys

$$\log_{10} 10^t = t. \tag{4.7}$$

There are a few natural processes that increase logarithmically in time, such as static friction between two objects in contact, but they are not common. The logarithm has two types of significance.

First, it provides a very handy calculational tool, because it converts multiplication to addition. That is

$$\ln[ab] = \ln[a] + \ln[b]. \tag{4.8}$$

Multiplication of a and b can therefore be carried out in the following way.

1. Find the logarithms $\ln a$ and $\ln b$ of a and b. This can be done by looking them up in tables.
2. Add the logarithms $\ln a + \ln b = \ln[ab]$.
3. Find the antilogarithm $ab = \exp[\ln ab]$.

This procedure was used for centuries, and underlies the construction of the *slide rule*, which was the mechanical device that preceded today's calculators.

Second, logarithms can be used to compactly represent physical measurements that range over many orders of magnitude. The physical units for quantities that vary greatly are based on logarithmic scales. For example, the decibel is

$$\text{Sound intensity in decibels} = 10 \log_{10} I/I_0$$

where I is the intensity of the sound in watts/m^2 and I_0 is a low sound intensity that corresponds roughly to the human threshold of hearing. So the way this scale works is that when some sound has 10 **times** more energy per area than another, its intensity in decibels equals that of the first **plus** 10. Similarly, the magnitude of earthquakes is defined in terms of the logarithm of the energy they release.

Sines and cosines   Sines and cosines have two basic uses in modeling.

First, sines and cosines give *projections*. For example, if a two-meter ladder leans against a wall at 70°, and the Sun is directly overhead, then the shadow cast by the ladder is $2 \cos 70°$ m long and the ladder rises to a height on the wall of $2 \sin 70°$ m.

Second, they are the functions to think about when faced with any periodic or recurring phenomenon. For this purpose, there is little difference between the two; the sine function starts at 0 ($\sin 0 = 0$) while the cosine function starts at 1 ($\cos 0 = 1$). The two functions are related to each other by a horizontal offset: $\sin(x) = \cos(x - \pi/2)$. Examples of when to think about these functions:

- Mean monthly temperature in Arizona as a function of month over 10 years.
- Air pressure measured every millisecond as a sound wave passes by.
- Sea level at a beach measured every 10 minutes over several days.

4.2 Estimation

4.2.1 Order of magnitude estimates

The physicist Enrico Fermi was famous for his ability to estimate almost anything. He was present at the explosion of the first atomic bomb in the New Mexico desert, and estimated the strength of the explosion by dropping pieces of paper and watching how far the blast from the bomb made them move as they fell. He demonstrated throughout his life that by making a series of reasonable assumptions, one can come up with a rough quantitative *estimate* for almost any question.

In a famous and rather useless example, he estimated the number of piano tuners in Chicago. Here is how an estimate like this can be done

1. Estimate the number of people in Chicago. Chicago is a large American city, and large American cities have populations of several million, so estimate 5 million.
2. Estimate the number of homes in Chicago. Here a guess would be that there are on average three times fewer homes than there are people, since most homes have families with two or more people in them.
3. Estimate the fraction of homes with a piano. Guess one in 10. There are also pianos at schools and theaters, but this number should be small by comparison, since schools and theaters serve hundreds or thousands of people and have just a few pianos.
4. Estimate the number of times per year pianos are tuned. Here, guess once per year.
5. Assume that the number of piano tuners has adjusted itself over time so that piano tuners have fairly steady business. If there were too many, they would not have work and quit, while if there were too few, demand for their services would raise the cost and more would enter the profession.
6. Estimate the time it takes per day to tune a piano. Guess 2 hours with transit time.
7. Estimate the number of hours per day a piano tuner works. Guess 8.
8. Estimate the number of days per year a piano tuner works. Guess 5 days a week times 50 weeks a year or 250.
9. Now, put everything together. First, find the number of piano tunings starting with the number of people. A way to keep track of the algebra is to set up a chart with two rows to contain all the guesses. Quantities that show up in the numerator of a guess are in the top row, and quantities in the denominator of a guess are in the bottom row. Make sure to keep track of the units of every guess. For example, saying that there is one house for every three people means that an ingredient in the final answer will be

$$\frac{1 \text{ home}}{3 \text{ people}}$$

The guesses in steps 1 through 4 assembled in a chart look as follows:

5 million people	1 home	1 piano	1 tuning
	3 people	20 homes	year

You will know that you have placed the entries correctly when the units cancel between the top and bottom rows. For example, the "homes" cancel between the second and third columns (it does not matter that "homes" is plural in the third column and singular in the second), since it is on top in the second column and on the bottom in the third. The answer at this stage is

$$\frac{5 \times 10^6 \text{ piano tuning}}{60 \text{ year}} = 83,000 \text{ piano tunings/year.}$$

10. Next find the total number of piano tunings per piano tuner in a year. It is

1	piano tuning	8 hours	250 days
piano tuner	2 hours	1 day	year

This gives

$$\frac{\text{piano tunings}}{\text{piano tuner}} \frac{8 \times 250}{2 \times \text{year}} = 1000 \text{ piano tunings/piano tuner/year.}$$

11. Finally, dividing Equation 9 by Equation 10 gives 83 tuners.

Some comments on this process are in order. Note that by keeping track of units, one can avoid the most common error, which is to multiply by numbers where one should divide or vice versa. In step 9 it would have been easy to multiply by 3 people/home rather than dividing. However in that case there would have been two factors of "people" in the numerator and two factors of "home" in the denominator when one tried to simplify the answer, making it possible to catch the error.

Next, since the final answer is made up of a series of guesses, the final answer is only as good as the guesses. However, the answers to smaller guesses such as "How many people are there in Chicago?" tend to be much more accurate than initial blind guesses to questions such as "How many piano tuners are there in Chicago." In addition, some of the guesses will tend to be high, others will tend to be low, and some of the error may cancel out this way. Finally, and probably most important, a series of guesses like this points to a process that can be refined. For example, one can actually look up the number of people in Chicago; the number in the city itself is around 3 million, and in the greater metropolitan area it is around 10 million.

The final guess does not seem to be all that bad. According to the US Bureau of Labor Statistics there are nearly 5000 musical instrument repairers and tuners in the U.S. Since Chicago has around 1/50th of the population of the US, this gives an estimate of 100 people repairing and tuning musical instruments around Chicago, which is consistent with an estimate of 80 piano tuners.

4.2.2 Dimensional analysis

The mathematics behind order of magnitude estimates is able to accomplish more than simply estimating numerical values. The final result is the product of intermediate quantities, each of which is easier to guess than the final answer by itself. It is easier to estimate the population of Chicago than to jump immediately to the number of piano tuners. Instead of using numerical values for intermediate values such as the population of Chicago, one can use symbols to represent them. In this case, the end result is a formula showing how some final result depends upon ingredients that make it up. Sometimes, thinking about cause and effect and paying attention to the units of quantities that can enter into some process is all that is needed to find a formula. This process is called *dimensional analysis* (Bridgman, 1921; Barenblatt, 1987) and when it works it seems almost like magic. There are limits to the magic. Formulas obtained in this way are always uncertain up to a dimensionless factor, such as a factor of 2 or π, but one can learn a lot from them nevertheless.

As a first example, consider a string on a guitar, and search for a formula giving the frequency f of a note when one plucks an open string. What can enter into the formula? The end result is a frequency, which has units of inverse time, s^{-1}. The note depends upon how tightly the string is pulled; when tuning it, one pulls on it and changes the tension. Tension T is a force, with units of kg $\times$ m/s^2. The final answer involves only units of inverse time, while tension also involves mass and length. The only mass that makes sense to employ is the mass M of the string. For a length, one could either use the length L of the string or its thickness t. If you play the guitar, you will know that you change the notes by pressing your fingers in front of frets and changing the length of the string that is able to vibrate. So it should be the length L of the string that matters, not the thickness. Dividing T by L and by M gives a quantity with dimensions of s^{-2}, and taking the square root gives something with units of inverse time. These observations suggest the formula for the frequency of a guitar string

$$f = C\sqrt{\frac{T}{ML}}, \qquad (4.9)$$

where C is a dimensionless constant. A detailed analysis based on equations of physics for wave motion in strings gives exactly the same result, with the additional information that the constant C is 1/2,

$$f = \frac{1}{2}\sqrt{\frac{T}{ML}}. \qquad (4.10)$$

Dimensional analysis by itself gives almost all the information that otherwise requires a much more lengthy and detailed analysis.

As a second example, suppose one has a ball flying through the air and wants to know the frictional force on it due to air resistance. Dimensional analysis can be used to find this too. In contrast with the frequency of a guitar string, there really is no method other than dimensional analysis that gives anything approaching a simple answer to this problem.

The quantity one wishes to find is a force F, which has units of kg $\times$ m/s^2. A formal way to indicate dimensions is by enclosing the force F in square brackets, where the brackets denote the dimensions of the quantity inside:

$$[F] = \text{kg} \times \text{m/s}^2. \tag{4.11}$$

What factors can affect the frictional force? The shape of the ball certainly plays a role, and the most reasonable feature of the shape is the cross-sectional area A that the ball presents to the wind. The speed of the ball v also must enter. The final answer involves units of mass, and neither the area of the ball nor the speed of the ball involves mass. So the mass must come either from the ball or from the air.

Deciding whether the mass of the ball or the air is involved requires a good understanding of the concept of force. The force of air on the ball is completely due to air molecules pushing on the outside of the ball as the ball and air molecules collide. How hard they push does not depend upon what is inside the ball. If it is full of heavy material, or hollow, the push is just the same. A similar example would arise if you were pushing as hard as you can on a car initially at rest; the force you apply depends on how strong you are, and on how well your feet grip the ground, not on whether the car is empty or loaded. The upshot of this argument is that the force the air applies to the ball depends on the shape of the ball, but not the mass, since the mass of the ball can change as much as one wishes by filling the ball with things with which the air cannot interact.

The only possibility left is that the force depends on the mass of the air. This at first seems to make no sense, since if the ball is moving in a small room, the total mass of air is much less than if it moves in a large room, but the friction force on the ball should not depend on the size of the room. This problem is solved if one divides the total mass M of air by the total volume V it occupies to obtain the density of air

$$\rho \equiv \frac{M}{V}. \tag{4.12}$$

It makes sense that frictional resistance should depend on this quantity. If one tries to throw a ball underwater it will not go nearly as far as in air, and this can be explained by noting that water is much more dense than air.

Assembling the quantities on which friction should depend, one has

$$[A] = \text{m}^2, \quad [v] = \text{m/s}, \quad [\rho] = \text{kg/m}^3. \tag{4.13}$$

The idea of dimensional analysis is that the desired result can be obtained as a product of powers. To express this idea formally, write

$$[F] = [A]^a [v]^b [\rho]^c; \tag{4.14}$$

that is, the units of force come from products of powers of the units of the input factors. Writing out Equation (4.14) gives

$$\text{kg m/s}^2 = \text{m}^{2a} (\text{m/s})^b (\text{kg/m}^3)^c. \tag{4.15}$$

While one can set up this equation as three simultaneous algebraic equations and solve them, the solution is easiest to obtain by inspection. The only way to get the powers of kg right is if $c = 1$. The only way to get powers of seconds, s, right is if $b = 2$. What remains is to get meters, m, correct. The powers of meters on the right-hand side are $2a + b - 3c = 2a + 2 - 3 = 2a - 1$. This has to equal 1, which is the correct power of meters on the left-hand side, so $a = 1$. Having determined a, b, and c, the formula for frictional force is

$$F = CA\rho v^2, \tag{4.16}$$

where C is a dimensionless constant. In this context, the constant C has a name, the *drag coefficient*. Drag coefficients are not at all easy to calculate, so they are usually obtained from experimental measurements, but they are constants of order one.

4.3 Linear regression

There are many experiments where a scientist gathers data, but it is not possible to repeat any of the measurements for precisely the same value of the independent variable. Many observational projects are like this. For example, suppose you obtain data on the strength of hurricanes versus the mean sea temperature on the day the hurricane has maximum amplitude. Different hurricanes will come into existence on days where the sea has different temperatures, so the data will explore a range of conditions. But there is no way to repeat the experiment. It is impossible to order up a hurricane at a desired (or undesired) spot with the sea at a prescribed temperature. You have to take what you get. Observational astronomy often presents a similar situation.

The statistical tools introduced so far do not quite work in these cases. The tools relied on the assumption that an experiment could be repeated as often as desired for a given value of the independent variable. But what can one do if with each new measurement there is a new value of the independent variable?

One possibility is to decide that when two measurements have independent variables that are close enough together, they are basically the same. This means

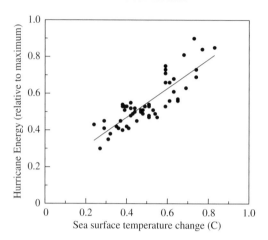

Figure 4.5 Scatter plot of annual average hurricane intensity versus annual mean surface temperature. The straight line is the best linear fit to the data. [Data from Emanuel, 2005.]

putting data into bins. This way of treating data is very general, because it involves no assumption about their form.

However the most common solution in such cases proceeds in a different way. Return to the example of hurricane intensity versus mean annual sea surface temperature. The first step in analyzing such data is to create a scatter plot, as shown in Figure 4.5. On inspection, it appears that while the data are scattered, they essentially trace out a straight line. Therefore, it makes sense to try to find the straight line that fits the data the most closely, and to estimate the certainty with which the slope and intercept of the line can be determined from the data. The best-fit line is called a *linear regression.*

The method of *least squares* finds this best-fit line. Here is how it works. Suppose you have N pairs of points, $(x_1, y_1), \ldots, (x_N, y_N)$. In Figure 4.5 the x-coordinate of each point is the sea surface temperatures and the y-coordinate is the hurricane intensity. Suppose that the best-fit line were known. It would be given by intercept and slope A and B. For any given x_i, the line would make the prediction

$$Y_i = A + Bx_i. \tag{4.17}$$

Unless all the points (x_i, y_i) really lie on a line, the points Y_i will not be the same as the points y_i; instead, the points (x_i, Y_i) all lie on the best-fit line.

How can one choose the intercept and slope A and B so as really to produce the best line? If A and B are the best coefficients, then the difference between the real points y_i and the estimates Y_i will be as small as possible. Since the differences between y_i and Y_i can be either positive or negative, it is not a good idea to form the sum $\sum_i (y_i - Y_i)$. The individual terms in the sum could be huge but cancel out due to sign changes. A better quantity to make as small as possible is the deviation D where

$$D(A, B) = \sum_{i=1}^{N}(y_i - Y_i)^2 = \sum_{i}^{N}(y_i - A - Bx_i)^2. \qquad (4.18)$$

When D has become as small as it can be by varying A and B, it must obey the two conditions

$$\frac{\partial D}{\partial A} = 0, \quad \frac{\partial D}{\partial B} = 0. \qquad (4.19)$$

Forming these two equations and carrying out some algebra gives

$$B = r\frac{s_y}{s_x} \qquad (4.20)$$

$$A = \bar{y} - B\bar{x}. \qquad (4.21)$$

Here, D is viewed as a function of A and B, and one uses the general rule from calculus that the derivative of a differentiable function vanishes at the minimum point. What are the quantities appearing in these equations? First, just as in Section 3.2.1, $\bar{x}$ and $\bar{y}$ are the sample averages of the sequences x_i and y_i:

$$\bar{x} = \sum_{i=1}^{N} x_i/N; \quad \bar{y} = \sum_{i=1}^{N} y_i/N. \qquad (4.22)$$

Next, as in Section 3.2.4, s_x and s_y are the sample standard deviations of the sequences x_i and y_i:

$$s_x = \sqrt{\sum_{i=1}^{N} \frac{(x_i - \bar{x})^2}{N - 1}}; \quad s_y = \sqrt{\sum_{i=1}^{N} \frac{(y_i - \bar{y})^2}{N - 1}}. \qquad (4.23)$$

Finally r is a new quantity called the *correlation coefficient*. It is defined to be

$$r = \sum_{i=1}^{N} \frac{1}{N - 1} \frac{(x_i - \bar{x})(y_i - \bar{y})}{s_x s_y}. \qquad (4.24)$$

The correlation coefficient r provides a measure of how well or poorly the points $(x_i, , y_i)$ fall on a straight line. When they fall perfectly on a line, then $r = 1$ if the line has positive slope, and $r = -1$ if the line has negative slope. If the points (x_i, y_i) do not perfectly fall on a line, then r lies somewhere between -1 and 1. If r is close to zero, one says that x and y are not linearly correlated, while if r is close to -1 or 1, one says that they are highly correlated, or well correlated, that a linear fit is excellent, or something of that sort.

For the hurricane data in Figure 4.5, the correlation coefficient is 0.85, which is a reasonably high level of correlation. Computing A and B from Equations (4.20) and (4.21) gives $A = 0.16$, and $B = 0.79/C$ (hurricane energy is measured in dimensionless units, so A is dimensionless and B has units of inverse temperature). The straight line in the figure is a plot of $Y = A + Bx$.

4.3.1 Error estimates for slope and intercept, and probability that the slope is really zero

Often the slope and intercept of a regression line have enough importance that one wants not only to measure them, but to find the uncertainty in the slope and intercept as well. The uncertainties can be obtained from the following expressions:

$$\Delta A = \frac{\sqrt{\frac{1}{N}\sum_{i=1}^{N}(y_i - A - Bx_i)^2}\sqrt{\frac{1}{N}\sum_i x_i^2}}{s_x\sqrt{N}} \tag{4.25}$$

$$\Delta B = \frac{\sqrt{\sum_{i=1}^{N}\frac{1}{N}(y_i - A - Bx_i)^2}}{s_x\sqrt{N}}. \tag{4.26}$$

Another question that can arise comes when the slope B is small, and the error bar from the estimate (4.26) overlaps or nearly overlaps 0. In this case, you might want to know whether the data really do have a slope, or whether you should adopt a null hypothesis that the slope is zero, and the function you have found is constant. To address this question, calculate the correlation coefficient r from Equation (4.24), and compute

$$t = r/\sqrt{(1 - r^2)/(N - 2)}. \tag{4.27}$$

To find the probability slope B was nonzero by chance alone, look up the probability of t in Table 3.6 on page 94 using $N - 2$ degrees of freedom.

4.4 Matching arbitrary functions to data

The techniques used in order to match linear functions to data can be used to match more complicated functions to data. The functions can be described by parameters $A, B, C, D, \ldots$ and one can repeat the whole process of finding the constants that make the difference between the function and observations as small as possible. There is a special case of these ideas that can be applied to any function. This is the case where your data have a shape that you recognize because it reminds you of a standard function. However, you still have the task of moving the standard function back and forth and expanding it in the vertical and horizontal directions so that it lies on top of your data. The basic mathematical operations are these:

- Multiplying a function $f(x)$ by a constant C makes it increase in scale in the *vertical* direction.
- Dividing the argument of a function $f(x)$ by a constant C, obtaining $f(x/C)$ makes the function increase in scale by factor C in the *horizontal* direction.
- Adding a constant as in $f(x) + C$ slides a function *up* by C.
- Subtracting a constant as in $f(x - C)$ slides a function *to the right* by C.

In short, these operations capture the ideas of *scale* and *offset*.

The need to understand scale and offset arises so often during inquiries that it requires some more detailed explanation. Here is an example that is intended to give you an idea of what to do in general. Rather than providing mathematical formulas to optimize scales and offsets, the discussion will focus on how to get a decent fit by eye.

Suppose you have done an inquiry where you have a turning wheel and you want to know how fast it is turning. You decide to measure this by putting a small speaker near the edge of the wheel, hooking it to a tone generator so that it plays a constant tone, and mounting a microphone on the wheel. As the wheel turns, the microphone gets nearer and farther from the speaker, and the strength of the sound it measures goes up and down. A schematic diagram of the apparatus appears in Figure 4.6 and a graph of sample data appears in Figure 4.7. As you are designing your experiment, the teaching assistant comes by and says that the data will be "basically sinusoidal."

You are not entirely clear on what this means, but the most logical guess is that the data are the same as $\sin(x)$. So you plot $\sin(x)$ (Figure 4.8). It looks nothing like the data. What was the TA talking about?

There are many reasons why the sine function does not look exactly like the data. There is nothing that can be done to make the fit perfect. However, the data and the function can be brought into much closer accord than seems apparent at first. In order to see how, it is valuable to have a computing environment that makes it possible to plot the data, type in a function, and see the function in comparison

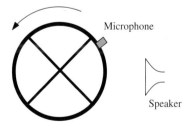

Figure 4.6 A microphone is attached to a rotating wheel, and catches sound from a speaker. The measured sound amplitude is used to find the rotation rate of the wheel.

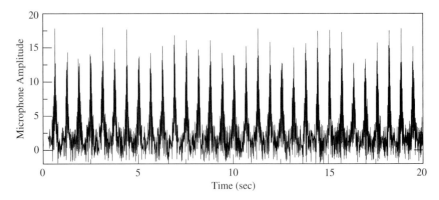

Figure 4.7 Here is a graph of the data collected by LoggerPro of output from the microphone versus time.

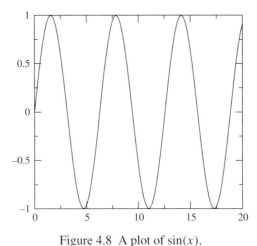

Figure 4.8 A plot of $\sin(x)$.

with the data. Any spreadsheet program will do, or you may prefer a specialized plotting program.

To begin the process of analyzing the data, zoom in on the region between 1 and 5 seconds (Figure 4.9). There are two oscillations visible. The slower oscillation takes about 1 second, which is more or less the distance between the highest peaks. This oscillation corresponds to the rotation of the wheel. How does one know? Well, from watching the wheel, that seems about right. The second oscillation is much more rapid, and comes from the rapid vibrations in air pressure measured by the microphone that our ears interpret as a tone. This assumption can be checked by turning on the speaker when the wheel is stationary and seeing that the fast oscillations are present but the slow oscillations are gone.

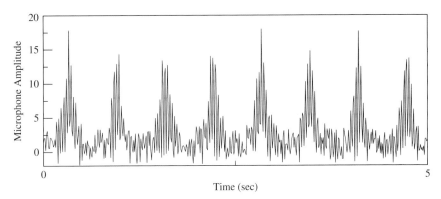

Figure 4.9 Zoom in on region between 1 and 5 seconds.

Since this project is aimed at measuring properties of the rotating wheel, and not the vibrations emitted by the speaker, further analysis focuses on the slow oscillation, neglecting the rapid one. It would be nice somehow to see the slow oscillation without the distracting rapid one getting in the way. This step is not absolutely necessary, and not every plotting program makes it easy, but it is not hard to automate in a spreadsheet program. The operation is to take a *running average*, which means one moves along the data, takes the average of, say, the first 10 points in the data. Then one collects together points 2 through 11, and takes the average of those, then points 3 through 12, and so on until one has traversed the whole data set. Averages have the effect of smoothing out wiggles, and a running average of 10 on these data smooths out any wiggles that have a wavelength smaller than 10, while leaving intact the structure that is much larger than 10 data points in width. Having performed this running average, the data now look as in Figure 4.10.

The analysis is now at the point where one can ask the questions about scale and offset with which the section began. The data look roughly like a sine wave, but in many details they are wrong. A sine function has a maximum at 1 and a minimum at -1. These data have a maximum at around 7 and a minimum at around 0. A sine function goes through a complete period when its argument goes through $2\pi \approx 6.2$. The data go through a complete period when time goes through 0.6 seconds, as one can find by measuring the distance between peaks. The sine function is zero when its argument is zero, while the data start at a value of 0.3. These four discrepancies between the data and the function are what can be cured with the four constants that govern scale and offset.

The vertical offset of a function is determined by adding a constant, and the vertical scale is determined by multiplying it by a constant. So the sine function is offset and scaled by writing

$$A + B \sin(x). \tag{4.28}$$

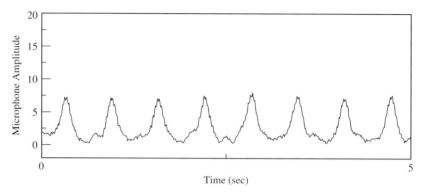

Figure 4.10 Take a running average over 10 consecutive points.

Choose A and B to match the maximum and minimum of the data. The sine function maximum at 1 should correspond to the data maximum at 7. The sine function minimum at -1 should correspond to the data minimum at 0. Write down the two relations

$$7 = A + B \times 1 \tag{4.29}$$

$$0 = A + B \times (-1) \tag{4.30}$$

$$\Rightarrow A = B = 3.5. \tag{4.31}$$

At this point one has the function

$$3.5 + 3.5 \sin(x). \tag{4.32}$$

Next, adjust the horizontal scale and offset. This is accomplished by applying addition and multiplication of constants to the argument of the sine function. That is, one considers the function

$$3.5 + 3.5 \sin(t/C - D). \tag{4.33}$$

The constants have a more natural interpretation if one divides and subtracts rather than adding and multiplying. To set the constant C, notice that when x travels through 0.6 sec, the sine function must travel through a full period, which means that x/C must change by 2π, or

$$2\pi = 0.6 \text{ sec}/C \Rightarrow C = 0.6 \text{ sec}/(2\pi) \approx 0.1 \text{ sec}. \tag{4.34}$$

One says that the period of the oscillation is 0.1 sec. The final constant to determine is D. The data have a first maximum when time equals 0.35 sec, while the sine function has a first maximum when its argument equals $\pi/2$. So

$$0.35\text{sec}/C - D = \pi/2 \Rightarrow D = 0.35\text{sec}/C - \pi/2 \approx 3.5 - 1.5 = 2. \tag{4.35}$$

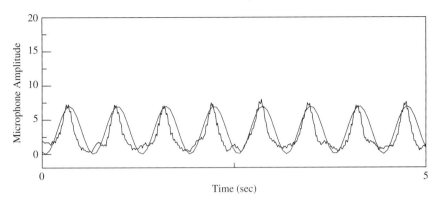

Figure 4.11 Data compared with properly scaled sine function.

Finally, one has the function

$$3.5 + 3.5\sin(t/(0.1\text{sec}) - 2). \tag{4.36}$$

A comparison of the scaled sine function with the data appears in Figure 4.11. Whether this is considered a good fit or not depends completely upon the context. A pure mathematician would be likely to think that the functions have nothing to do with one another and not worth further discussion. An applied mathematician would want to find a way to quantify the difference between the data and the fit. A theoretical physicist would view the situation as an encouraging start, but would suggest 15 ways to find more accurate functions, all of them time consuming (and requiring extensive summer funding.) An experimental physicist would view the theoretical fit as acceptable, but want to improve the quality of the data (requiring extensive summer funding). A biologist would view the fit as perfect.

4.5 Fourier transforms

The Fourier transform is a powerful tool for analyzing data that vary continuously in time or in space. It is most useful for signals that contain a repeating pattern, like the data in Figure 4.11. The basis of the tool is a remarkable theorem which states that any continuous function can be rewritten exactly and uniquely as a sum of sine and cosine functions. When the continuous function contains repeating patterns, the Fourier transform homes in on the repeating parts, describes their frequencies, and also their amplitudes.

You do not have to know the theorem in order to make use of the Fourier transform, but seeing it spelled out should help interpret the results. The theorem says this: suppose you have a function $F(t)$ defined on the time interval $[0, T]$. Then

$F(t)$ can be rewritten as

$$F(t) = A_0 + A_1 \cos(2\pi t/T) + A_2 \cos(4\pi t/T) + A_3 \cos(6\pi t/T) + \ldots$$
$$+ B_1 \sin(2\pi t/T) + B_2 \sin(4\pi t/T) + B_3 \sin(6\pi t/T) + \ldots \quad (4.37)$$

The cosine and sine functions that appear in the sum are all the cosine and sine functions that are periodic over the interval T. The first one that shows up, $\cos(2\pi t/T)$ or $\sin(2\pi t/T)$, has one complete period in the interval $[0, T]$. The second one that shows up has two complete periods in the interval $[0, T]$, the third one three complete periods. The coefficients A_n and B_n are uniquely determined by the following expressions

$$A_0 = \frac{1}{T} \int_0^T F(t)\, dt$$

$$A_1 = \frac{2}{T} \int_0^T F(t) \cos(2\pi t/T)\, dt$$

$$A_2 = \frac{2}{T} \int_0^T F(t) \cos(4\pi t/T)\, dt$$

$$A_3 = \frac{2}{T} \int_0^T F(t) \cos(6\pi t/T)\, dt$$

$$\vdots$$

$$B_1 = \frac{2}{T} \int_0^T F(t) \sin(2\pi t/T)\, dt$$

$$B_2 = \frac{2}{T} \int_0^T F(t) \sin(4\pi t/T)\, dt$$

$$B_3 = \frac{2}{T} \int_0^T F(t) \sin(6\pi t/T)\, dt$$

$$\vdots \quad (4.38)$$

Each of these integrals goes to the function $F(t)$, asks "How important is $\cos(n\pi t/T)$ to you?" and comes back with a number that says exactly how much of $\cos(n\pi t/T)$ or $\sin(n\pi t/T)$ needs to be put into a sum to get back the original function. The process of starting with the function $F(t)$ and ending with the coefficients A_n and B_n is called the Fourier transform, and A_n and B_n are called Fourier coefficients.

Unfortunately, there is no easy way to use spreadsheets to obtain the Fourier coefficients A_n and B_n. However, if you gather data with a program such as

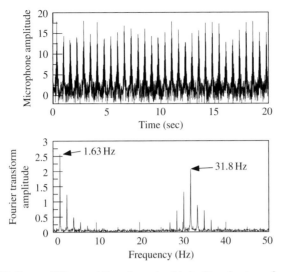

Figure 4.12 Data of Figure 4.7 replotted with its Fourier transform below.

LoggerPro, you can obtain them simply from pull-down menus. As an example, return to the data in Figure 4.7. Making use of the discrete Fourier transform in LoggerPro gives the plot shown in Figure 4.12.

What exactly is the numerical Fourier transform reporting? Rather than reporting the coefficients A_n and B_n separately, the plot shows them combined as $\sqrt{A_n^2 + B_n^2}$. What is on the horizontal axis of the Fourier transform plot? It is not the integer n that indexes A and B. Instead, the horizontal axis shows the frequency of oscillation f to which the coefficient corresponds, which is

$$f = \frac{n}{T}. \tag{4.39}$$

That is, the Fourier transform graph shows the amplitude of the coefficients of $\cos 2\pi ft$ and $\sin 2\pi ft$ as a function of f. Going back to Figure 4.11, the slow period of oscillation in the data is around 0.6 s. This corresponds to a frequency of $1/0.6\text{s}=1.66$ Hz. The Fourier transform immediately gives a more precise result. The largest peak is at $f = 1.63$ Hz. The horizontal coordinate of peaks gives the frequency of the most important components of the original signal. A second group of peaks is visible, centered at $f = 31.8$ Hz. This higher frequency is the frequency of the sound wave emitted by the speaker.

Inspection of the Fourier transform graph can raise more questions. Why is each of the two highest peaks surrounded by other peaks, evenly spaced and falling off in amplitude? Answers to questions such as these and much more can be found in texts with fuller discussion of Fourier transforms, such as Crawford (1968).

4.6 Deterministic modeling

Deterministic modeling is one of the great ideas in the history of science and mathematics. The idea is to use mathematics to predict the future. Every deterministic model has two basic elements. The first element is a set of equations that explain how some part of the world changes. The second element is precise knowledge about how that part of the world is at some point in time. The first element is called *equations of motion*, and the second is called *initial conditions*.

Once these elements are present, the model works like this. It starts with the initial conditions, and uses the equations of motion to find a new state of the world some time in the future. Once the calculations are done, a new state of the world is known, so the equations of motion can be used again to make a prediction even farther into the future. And again. And again. For as long as one has time to wait for the calculations.

It is virtually always true that as you try to move further and further into the future with mathematical equations describing change, your predictions will become less and less accurate. Weather forecasts are an example of deterministic modeling where the loss of accuracy is familiar. Weather forecasts are based on knowledge of the state of the weather right now, and use calculations to predict what will happen in the future. Forecasts within a single day are pretty good, as are those for one or two days into the future. Forecasts a week in advance are not very reliable. For some equations of motion, errors grow as a power law of the time into the future one is trying to project, while for others errors grow exponentially fast. Errors in weather prediction grow exponentially, which is why accurate weather forecasting cannot be carried out more than a few days into the future.

4.6.1 Iterative maps

Here is an example of modeling that comes from trying to predict the population of the United States. An accurate model is not easy to construct. The U.S. Census Bureau makes projections after breaking down population by state, region, gender, ethnic group, and accounting for births, deaths, immigration and emigration, migration from one part of the United States to another, and movement to and from overseas of U.S. Armed Forces. For a first model, however, one can simply use the most natural idea about population changes, which is that change in population is proportional to the size of the population. The reason for this is that the more men and women there are in a population, the more children they will have. On the other hand, the larger a population, the more people will die each year as well. The natural increase of population is the surplus of births over deaths. If both births and deaths are proportional to the size of the population, then the difference between them will be proportional to the population as well. These thoughts lead to a very simple equation of motion for population,

$$X_{n+1} = X_n + rX_n. \tag{4.40}$$

Here X_n is the population of the United States in decade n, and X_{n+1} is the population in the next decade. The constant r gives the net increase in population per decade. If you know the population of the U.S. in decade 0, the equation gives you the population in decade 1. It is

$$X_1 = X_0 + rX_0. \tag{4.41}$$

But now, since you know the population in decade 1, you can find it in decade 2:

$$X_2 = X_1 + rX_1. \tag{4.42}$$

And so on forever. There is just one problem. The whole discussion is abstract, and in order to get numbers out, you need actually to know the initial condition – the population in decade 0 – and you need a value for the rate of increase per decade r. That is not so simple if you have to do all the work from scratch. The first U.S. census was carried out in 1790 under order of Thomas Jefferson by marshalls traveling the country on horseback. The U.S. Census Bureau today regularly employs around 12,000 people, but when the census was conducted in 2000 over 860,000 temporary workers were brought in. All that mainly (although not only) to get the result that in 2000 the population of the United States was 281,42,1906. Using historical records, the average increase in population of the United States per decade (US Census Bureau, 2008) from 1950 to 2000 was 14%, and one can use this value for r.

So the model is now complete. You can choose any decade to correspond to $n = 0$; for example, you can take $n = 0$ to correspond to the value of the 2000 census. Using Equation (4.40) to project the population forward, starting in the year 2000 and proceeding to 2050 gives the results shown in Figure 4.13.

Iterating a nonlinear map in a spreadsheet

A great thing about iterated maps is that it is easy to think up new rules for how things change and then test them. Suppose for example that there is some maximum population the U.S. can sustain, and you decide that population growth must slow down as it approaches the maximum. There are many possible mathematical rules that would produce this effect. A simple one is to guess that the growth rate r in Equation (4.40) should be replaced by

$$r \rightarrow r \times (1 - X_n/X_{\max}) \tag{4.43}$$

where $X_{\max}$ is the maximum possible population. With this replacement for the growth rate r the rate of population increase will slow down linearly as the population X_n approaches the maximum $X_{\max}$. A new equation for the population growth of the U.S. is

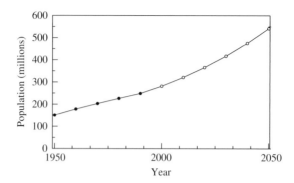

Figure 4.13 Population of the U.S. from 1950 to 2050, using census data up until 2000, and from then on employing Equation (4.40). Solid circles are census data, and hollow circles are projections. This is not a very reliable projection of population, just a first pass based on the simplest possible equation.

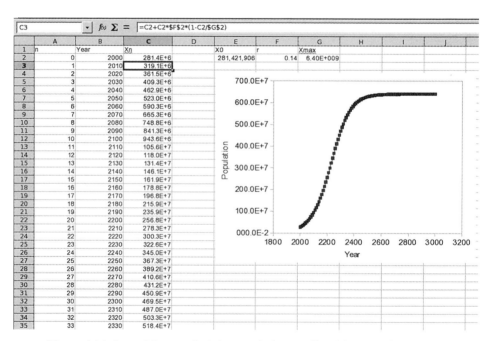

Figure 4.14 Spreadsheet to find the population predicted by Equation (4.44).

$$X_{n+1} = X_n + r(1 - X_n/X_{\max})X_n. \tag{4.44}$$

The right-hand side of this equation involves two powers of X_n; the equation is called *nonlinear* and it is not possible in general to write down a solution for X_n as a function of the starting value X_0. However, this sort of equation is very easy for computers to handle, and you can solve it in any spreadsheet. Figure 4.14 shows how to set things up. Column A contains the index n, which in this case will refer

Table 4.1 Population of U.S., 1950–2000. Source, US Census Bureau (2008).

Decade	U.S. population	Ratio to previous decade
1950	150,697,361	1.14
1960	178,464,236	1.18
1970	203,211,926	1.14
1980	226,545,805	1.11
1990	248,709,873	1.10
2000	281,421,906	1.13

to the decade, starting with the year 2000. So $n = 0$ will refer to the year 2000, $n = 1$ refers to 2010, and so on. To make this correspondence explicit, column B contains the year. One way to fill in the years is to write in cell B2 the formula =2000+A2*10 and then copy this formula into cells B3:B22. Column C is the one that does all the work. Cell C2 contains X_0, the starting population which is set to the U.S. census value from Table 4.1. This cell is different from all the ones below, because this one contains a number while all the others have a numerical representation of Equation (4.44). To emphasize this difference, the value in cell C2 comes from the expression =E2, copying over the value in cell E2. Now comes the really important cell, C3. This cell is highlighted and its contents show up in the formula bar above. It contains the expression

$$= C2 + C2 * \$F\$2 * (1 - C2/\$G\$2). \tag{4.45}$$

The correspondence between Equation (4.44) and C3 is

$$C2 = X_0$$
$$C3 = X_1$$
$$F2 = r$$
$$G2 = X_{\max}$$

so cell C3 corresponds to the equation

$$X_1 = X_0 + X_0 \times r \times (1 - X_0/X_{\max}). \tag{4.46}$$

To make the spreadsheet do as much work as possible, the goal is to copy the contents of cell C3 into cells C4:C102, and have successive rows automatically compute the iterated map in Equation (4.44). This works, just so long as one is careful about references to cells. There is just one value of r and just one value of $X_{\max}$ that all cells need to share. Therefore in cell C3, these cells are referred to by the absolute cell names $F$2 and $G$2. When C3 is copied down below, these references do not change. However, in cell C4, one would like to have the equation for X_2, which is

$$X_2 \quad = \quad X_1 + X_1 \times r \times (1 - X_1/X_{max})$$
$$\Rightarrow \text{C4 contains} = \text{C3} + \text{C3} * \$\text{F}\$2 * (1 - \text{C3}/\$\text{G}\$2).$$

But that is what spreadsheets do automatically. When you copy a formula into a cell that refers to a cell one above it, after copying, the formula continues to refer to the cell just above it, unless dollar signs tell the spreadsheet to act differently. So copying the contents of cell C3 into C4:C102 does in fact iterate the map (4.44), and the results appear both in numerical and graphical form. For the purposes of the spreadsheet, r is set to the average U.S. population growth per decade since 1950 of 14% per decade, and X_{max} is set to 6.4 billion people. Note that according to these calculations, the U.S. population will saturate at the maximum around the year 2400. The maximum population is a parameter put into the model, so the mathematics cannot be given either credit or blame for reporting it, and this value is not a prediction of the calculation. What the calculation does predict is the way that the population will behave over time in approaching the maximum. Even this prediction is only as good as the assumptions that led to Equation (4.44). But this approach to modeling is certainly more sophisticated than assuming, say, that the population of the U.S. will increase like a linear, quadratic, or exponential function.

4.6.2 *Differential equations*

One of the main purposes of calculus is to make it possible to obtain predictions about things that change continuously in time, using the same basic principles presented in Section 4.6.1 for iterative maps. The idea is still the same. You need two ingredients. First, you need an idea for how some system changes. Second, you need data telling you particular values for the system that interests you at some point in time. The idea for how the system changes is a set of *differential equations*. The data on particular values are *initial conditions*. You take the initial conditions, integrate them forward in time, and obtain predictions. This may sound much fancier than the population modeling of the previous section, but it is in fact almost exactly the same.

As an example, consider an experiment where you place a metal temperature probe into a glass of hot water and measure the temperature over time (Figure 4.15). Just based on common experience, you should know that heat flows from hot objects to cold ones, and when two objects are at the same temperature, heat does not move between them. This observation leads to a guess about the temperature of the probe. Suppose the temperature of the probe is T_p and the temperature of the

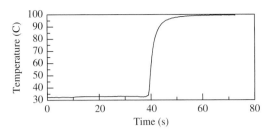

Figure 4.15 Data from a temperature probe placed in a beaker of boiling water as a function of time.

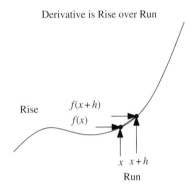

Figure 4.16 The definition of the derivative comes from computing the slope of a straight line drawn through nearby points of an arbitrary function, $f(x)$.

water is T_w. The temperature of the probe should stop changing when it reaches the temperature of the water, and the greater the difference in temperature, the faster it should change. This leads to an idea that

$$\text{rate of change of probe temperature} = K(T_w - T_p). \qquad (4.47)$$

Here K is some constant with units of inverse time. The constant K needs to be positive so that the probe temperature T_p increases when it is less than the water temperature. The time rate of change of the probe temperature is by definition the derivative with respect to time, dT_p/dt, so

$$\frac{dT_p}{dt} = K(T_w - T_p). \qquad (4.48)$$

This might be a good place to make some comments about the basic idea of differential calculus, and notation used for it. The basic idea is illustrated in Figure 4.16. Suppose you have any function $f(x)$. You want to know the rate at which the function is changing at some point x. The way to do this is to choose a second point $x + h$ that is a little bit to the right of the first point. The value of the function f increases from $f(x)$ to $f(x + h)$. So if you draw a straight line between these two points, the second point lies $f(x + h) - f(x)$ above the first, and a distance h to the right of it. The rise over the run for this straight line is

$$\text{Slope} = \frac{f(x + h) - f(x)}{h}. \tag{4.49}$$

The problem of notation

One of the problems in science is that there are many more than 100 ideas and quantities for which one would like symbols. This means that the Latin and Greek alphabets combined cannot provide enough characters to give every quantity a unique symbol that everyone agrees will refer to it forever.

In every scientific community, scientists become used to using symbols certain ways. For statisticians t brings up the idea of the t statistic (Section 3.7) while for physicists it brings up the idea of time. There are great advantages to using familiar symbols. It is much easier to understand an equation when the meaning of most of the symbols in it is obvious. Thus the slope of the function $y(x)$ at point x is often written as

$$y'(x) = \frac{dy}{dx}. \tag{4.50}$$

People who have had calculus know that the symbol d is special. You cannot divide through by d as you would in an ordinary fraction. Here at least is a convention that has been adopted universally. When you see a fraction with d on the left-hand side top and bottom, it's a derivative and the d is special. But there are pitfalls as well. Sometimes an equation is meant to represent an idea, and students think the idea applies only when those specific symbols are used. So, while d is special in the derivative, y is not. The derivative can find the slope of **any** function with respect to **any** variable. The rate at which concentration $c(t)$ changes in time is

$$c'(t) = \frac{dc}{dt}. \tag{4.51}$$

So if you look at this expression and find yourself asking "Is c the same as y?" then you have been misled by the fact that some math classes use only y or f to represent functions. Scientists take derivatives of functions that use every letter in all alphabets, and the idea of the derivative does not change because y is replaced by c.

One solution to the problem of running out of symbols is to use subscripts. If you are doing experiments with collisions of a heavy and a light car, you could use M_h and M_l to describe their masses. The notation is compact and helps you remember what you are referring to when you start doing algebra.

As the distance h gets smaller and smaller, the line connecting these two points gets closer and closer to a true tangent line to the curve. When h is small enough, Equation (4.49) is a good approximation, so one can write

$$f'(x) = \frac{df}{dx} \approx \frac{f(x+h) - f(x)}{h}. \qquad (4.52)$$

This is a somewhat flexible standard. You could decide that an answer is close enough to the right one to satisfy you when it is within 0.1%. Then you just have to choose h small enough to get this desired accuracy. The notation $f'(x)$ and df/dx describe the same thing: the slope of the curve $f(x)$ at point x, or the rate at which the function $f(x)$ changes when x changes.

While all these statements should be familiar from calculus, some features of them need to be emphasized. The function f can be any function, and it does not need to be named f. In fact it also does not need to be named y. The constant h does not have to be named h. It can be named ϵ, or dx to denote a small change in x. So, choosing symbols for functions somewhat randomly,

$$q'(l) \approx \frac{q(l+h) - q(l)}{h} \qquad (4.53)$$

is (approximately) the derivative of q at l, and in the case of the temperature probe,

$$\frac{dT_p}{dt} \approx \frac{T_p(t+dt) - T_p(t)}{dt} \qquad (4.54)$$

is (approximately) the derivative with respect to t, or equivalently the time rate of change of the temperature probe. On the left-hand side of this expression, dt denotes taking the derivative of $T_p(t)$, while on the right-hand side it denotes a small change in time.

Courses in calculus and ordinary differential equations spend a lot of time presenting methods to solve differential equations such as Equation (4.48) exactly. These methods have many uses but really, they are *so twentieth century*. The modern way to approach equations like this is to let a computer do most of the work. And the way to do that is to take Equation (4.54) literally. Combining Equations (4.48) and (4.54), write

$$\frac{T_p(t+dt) - T_p(t)}{dt} \approx K(T_w - T_p) \qquad (4.55)$$

and solve for $T_p(t+dt)$ to get

$$T_p(t+dt) \approx T_p(t) + dt \times K(T_w - T_p(t)). \qquad (4.56)$$

Now is a good time to look back at Equations (4.40) and (4.44) because (4.56) has the same basic structure. The first item on the right-hand side is some quantity (population or temperature) that one can suppose to be known at some time. The next item on the right-hand side expresses how that quantity changes during some time interval. For the population models, the time interval was a decade. Here, the time interval is denoted by dt and its value has not yet been settled, but it has to be small enough.

> **When is the time interval dt small enough in Equation (4.56)?**
> In a calculus class, you would have to take the limit where dt goes to zero. For numerical solutions literally going to that limit is completely impractical. Instead, you need to make the time interval dt small enough. One way to think about it is that you make dt smaller and smaller until results you care about such as the temperature after 10 seconds stop changing anymore. Another way to think about it is that there is some natural time-scale on which quantities in your experiment change and you need to make dt something like 10 times smaller than that. When in doubt, reduce dt and check how the final results depend on it.

On the left-hand side of Equation (4.56) is the quantity (population or temperature) at the end of the time interval. And the wonderful thing about equations of this form is that they can be used iteratively, over and over, to obtain predictions as far as one would like into the future.

Figures 4.17–4.19 display a process of using a spreadsheet program in order to solve the recursion relation in Equation (4.56). The spreadsheets make use of relative references between cells, which means that most of the work can be done by getting one row right and then copying and pasting to create many more rows. Spreadsheets are not the preferred tool of scientists in solving differential

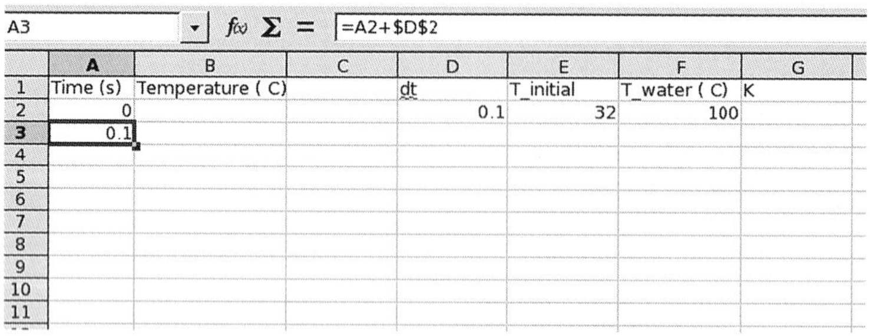

A3			$f\!x$ Σ =	=A2+$D$2			
	A	B	C	D	E	F	G
1	Time (s)	Temperature (C)		dt	T_initial	T_water (C)	K
2	0			0.1	32	100	
3	0.1						
4							
5							
6							
7							
8							
9							
10							
11							

Figure 4.17 Beginning of a spreadsheet to integrate predictions of Equation (4.56) into the future. The two most important columns are A which has the time and B which has the temperature. Parameters needed for the computation appear in cells D2 through G2. Note that the spacing between consecutive times is given by the contents of cell D2, and therefore updates automatically when D2 changes.

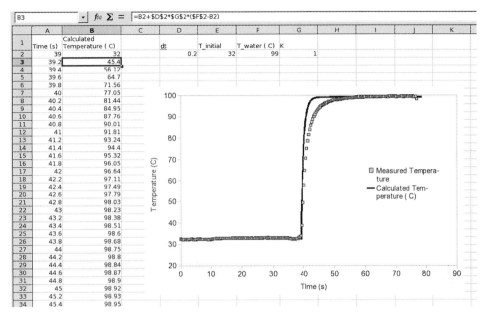

Figure 4.18 The spreadsheet is now largely completed. The starting time has been changed from 0 to 39 s to match the data. Equation (4.56) has been entered in cell B2 and copied into cells B3:B200. The experimental data are plotted alongside the predictions. An arbitrary value $K = 1$ C/s has been entered and theory and experiment do not yet agree very well.

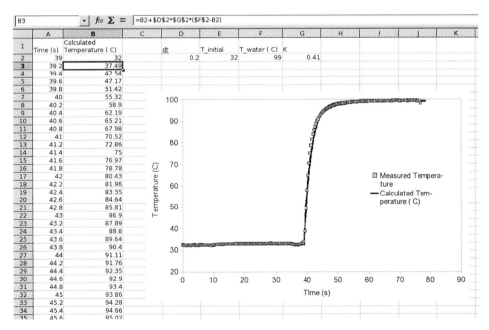

Figure 4.19 Through trial and error, K is adjusted to value 0.41 C/s. To the eye, comparison between theory and experiment is now acceptable.

equations, because they compute relatively slowly, and it becomes cumbersome or even impossible to use them if one wants to iterate relations millions of times. More sophisticated methods can be found in books such as Gershenfeld (1999) or Press *et al.* (2007). However, for the purpose of understanding the basic idea behind numerical solutions of differential equations and recursion relations, it is hard to find a simpler way to get numbers and pictures in a hurry.

Assignments

4.1 **Basic operations** Solve each of the following word problems, explicitly pointing out the role of addition, subtraction, multiplication, and division in finding each solution.

   a. My car is traveling at 20 miles per hour. How many meters does it go after 10 seconds?

   b. Once he reached his twentieth birthday, John started paying $20 per year for physical exams. How old was he when he had paid $180?

   c. An internship program pays students $10/hour. Students work an average of 15 hours/week for 12 weeks, and typically 45 students participate per semester. How much money does the program cost per year?

   d. An internship program pays students $10/hour. Students work an average of 15 hours/week for 12 weeks. You have $100,000 for two semesters. How many students can you take into the program?

   e. On Monday my plant is 2 cm long. It grows 0.3 cm/day. How long is it after a week?

   f. My plant grows 0.3 cm/day. After a week it is 7 cm long. How long was it originally?

   g. On Monday my plant is 2 cm long. After a week it is 7 cm long. How fast does it grow?

4.2 **Sketching and labeling**

   a. A student shines a flashlight directly towards an optical sensor, and measures the intensity of the light hitting it as a function of distance. The sensor is around half an inch in diameter, and she holds the flashlight at a distance ranging between 1 and 5 meters from the sensor. Light from the flashlight fans out from it in a cone.

     i. Draw a sketch of this experiment, and indicate in words all features that should be relevant in determining light intensity as a function of distance.

     ii. Label the sketch with symbols that could be used in order to construct mathematical equations relating light intensity and distance.

   b. A student goes into an elevator, puts a stand on the floor, attaches a force sensor to the top of the stand, hangs a spring from the force sensor, and attaches a 500 gram mass to the bottom of the spring, which stretches but does not come near the floor. The student measures the force on the sensor as the elevator goes up and down.

     i. Draw a sketch of this experiment, and indicate in words all features that should be relevant in determining the force measured by the force sensor.

ii. Label the sketch with symbols that could be used in order to construct mathematical equations describing the forces acting on the sensor as a function of time.

4.3 **Order of magnitude estimates** The purpose of this exercise is for you to get some practice with systematic thinking and estimation using tools and techniques you already know. You are not expected to come up with an answer that is precisely correct. Rather, your answer should be within an order of magnitude, or a factor of about 10, of the correct answer. We are looking to see how you think about a problem and go about logically and systematically arriving at an answer. Order of magnitude estimation is a tremendously useful skill for scientists designing an experiment or checking a theory, and more generally for a wide variety of decision making processes. It also provides you with a way to analyze situations that are inherently complex.

a. How many hairs does a typical person have on her head?
b. How much gasoline is consumed in the U.S.A. per year? Also estimate the amount of CO_2 produced as a byproduct? (Give your answers in gallons and tons.)
c. How much does the Eiffel tower weigh?
d. Estimate the total sales per year for a typical supermarket store in your home town (not an entire chain, just one store)? What is the net profit (before taxes) per year?
e. How many atoms are there in one lip's worth of lip balm?
f. How high can you jump on the Moon? (Moon's gravity is 1/6 of Earth's.)
g. How many bits are transmitted in an hour's worth of TV programming?
h. How many neurons do we have? How about a goldfish?

You may look up any number that you might need for your calculation such as the height and width of the Eiffel tower, or the mass of an atom. Obviously you get no credit for looking up the whole answer. You should provide enough details so that instructors can see how you arrived at your answer; you will receive no credit for just writing down a number even when that number is correct.

4.4 **Dimensional analysis** If a ball moves slowly enough through a thick liquid, such as molasses, then Equation (4.16) is wrong. Liquid flows smoothly around the ball in a regular pattern. The resistance the ball feels going through the liquid depends upon how easy it is to make the liquid flow past itself. This may be measured by putting the fluid between two flat plates, and moving the plates slowly with respect to one another. The property of the fluid is called *viscosity,* and is denoted by ν. It has units

$$[\nu] = \frac{\text{kg}}{\text{m s}}.$$

Use dimensional analysis to find a formula for the frictional force a ball experiences moving in a viscous liquid. The viscosity of the liquid appears in the formula, and the density of the liquid does not.

4.5 **M&M decay** A slice is drawn in a pie pan, 167 M&Ms are placed in the pan and shaken around randomly. All those that are in the slice are removed. 116 M&Ms remain. The remaining M&Ms are shaken randomly again, and again those remaining in the slice are removed. Data from this experiment are recorded below:

Trial	M&Ms left
0	167
1	116
2	82
3	56
4	40
5	27
6	20
7	13
8	10
9	7

   a. Find a function to fit the data. Please use a function that makes a sensible prediction about what happens as the number of trials goes to infinity.

   b. Calculate χ^2, and check whether your fit is acceptable. Note that the experiment counts the number of M&M's that belong to a category given by the trial number, so it is acceptable to use the form of the χ^2 test in Equation (3.76).

4.6 **Population modelling** The goal of this set of exercises is to illustrate the way that simple mathematics and computing can be used to develop models.

   a. The simplest model of population growth is

$$X_{n+1} = R_0 X_n, \tag{4.57}$$

where R_0 is the annual growth rate, and X_n is the population of some group of organisms in year number n.

    i. If $R_0 = 1.2$ for foxes, and there are 2000 foxes in year zero, how long is it before foxes weigh as much as the Earth?

    ii. If $R_0 = 0.5$, for foxes and there are 2000 foxes in year zero, what happens as time progresses?

    iii. Generalize your results. The behavior of this model divides into three qualitatively different regimes for low, middle, and high values of R_0. What are the values of R_0 that produce these three qualitatively different outcomes at long times and why?

   b. Something must be done to prevent uncontrolled growth at long times. Let the growth rate depend upon the population, and ask that the growth rate decrease when the population becomes too large

$$X_{n+1} = R(X_n) X_n. \tag{4.58}$$

The simplest possible decreasing function is

$$R(X_n) = R_0[1 - X_n/X_{max}] \tag{4.59}$$

where X_{max} is a population at which population drops immediately to zero. So now we have

$$X_{n+1} = R_0[1 - X_n/X_{max}]X_n. \tag{4.60}$$

   i. Define a new variable Y_n so that dependence upon X_{max} drops out, R_0 is the only free parameter of the model.
   ii. Express the meaning of Y_n in words.
 c. The *logistic map* is $Y_{n+1} = R_0[1 - Y_n]Y_n$.
   i. Set $R_0 = 0.5$. Starting with any population Y_0 between 0 and 1, what is the population after 30 iterations of the map?
   ii. What value is the population tending towards?
   iii. Why?
   iv. Set $R_0 = 1.2$. Starting with any population Y_0 between 0 and 1, what is the population after 30 iterations of the map?
   v. What value is the population tending towards? How does it depend upon the starting value? That is, let Y_∞ be the population reached after an infinite number of iterations. Find an expression for the function $Y_\infty(R_0)$. The value Y_∞ is called a *fixed point*.
   vi. Set $R_0 = 3.2$. Starting with any population Y_0 between 0 and 1, what is the population after 30 iterations of the map? Draw a chart of the results. What pattern is it tending towards?
   vii. Find through numerical experimentation the value of R_0 at which the map makes a transition between exhibiting a fixed point at long times, and the sort of behavior you saw in the previous part.
   viii. Explain this value analytically. Hint: Show that $Y_\infty(R_0)$ is a fixed point for all values of R_0. However, it is not *stable*. To understand the meaning of this claim, consider the evolution of $Y_n = Y_\infty + \delta_n$ under the action of the map, where $\delta_n \ll 1$. Find analytically the condition for δ_n to grow.
   ix. One must have $0 \leq R_0 \leq 4$. Why?
   x. Experiment with other values of R_0, just so long as $0 < R_0 < 4$. Carefully describe the long-time behavior of the model for values of R_0 just above 3.2. How many qualitatively different transitions can you find?

References

G. I. Barenblatt (1987), *Dimensional Analysis*, Gordon and Breach, New York.

P. W. Bridgman (1921), *Dimensional Analysis*, Yale University Press, New Haven.

F. S. Crawford (1968), *Waves*, McGraw-Hill, New York.

K. Emanuel (2005), Increasing destructiveness of tropical cyclones over the past 30 years, *Nature*, **436**, 686.

N. Gershenfeld (1999), *The Nature of Mathematical Modeling*, Cambridge University Press, Cambridge.

A. Martinez (2006), *Negative Math: How Mathematical Rules can be Positively Bent*, Princeton University Press, Princeton.

W. H. Press, S. A. Teukolsky, W. T. Vetterling, and B. P. Flannery (2007), *Numerical Recipes: The Art of Scientific Computing*, Cambridge University Press, Cambridge.

U.S. Census Bureau (2008), http://www.census.gov/population/www/documentation/twps0056.html

5

Scientific information

5.1 Introduction

If a scientist carries out a major research project, but no one knows about it, or no one can understand it, the research is of little use. So a large part of science is communication. The communication happens informally between colleagues in the hallway, at conferences through presentations, and most durably with global distribution of information through articles and books. Scientists have developed many conventional ways to explain their results, including equations, figures, and specialized vocabulary. Once you have completed a research project, you will need to practice communicating the results, to round out the set of scientific skills you have acquired. Each of the sections of this chapter touches just on a few essentials. You can find much more complete discussion of all the topics mentioned here in Valiela (2001).

5.2 Writing a proposal

Scientists spend a large fraction of their time writing proposals, whether they want to or not. Proposals are necessary to apply for grant funding, or for permission to conduct research. The research proposals needed for funding from federal agencies are usually a minimum of 15 pages in length, not including budgets, bibliography, and supporting documents. In this class you will not have to write proposals of that length. However, your instructors may ask you to write proposals a page or two long as you are beginning your inquiries.

The proposals will both help you focus your thoughts, and give the instructors a better opportunity to keep track of what you are doing and hence give you feedback. The more detailed your proposal, the better feedback you will get, and the stronger your final project will be. Here are a few guidelines for writing research proposals.

1. Chapter 1 lists a number of different types of research projects. Decide which type of project you are conducting. Usually projects Test a Hypothesis, Measure a Relationship, or (more rarely) Measure a Value or Construct a Model.
2. Explain briefly what motivates you to carry out this project.
3. Whether testing a hypothesis or measuring a relationship, you should clearly describe your dependent and independent variables. When testing an hypothesis, you should clearly state your null and alternative hypotheses.
4. Draw a made-up graph of what you think your data might look like when the experiment is complete. What type of data are your dependent and independent variables: Categorical? Continuous?
5. What type of data will you collect? How will you collect the data? What equipment do you need? How many measurements do you need to take? The more detailed your plans are, the better instructors can identify and head off problems before they arise. Ideally, you may want to write out step-by-step instructions, like writing a lab manual for your experiment.

You do not need to follow exactly this format, nor do all of these items need a lot of detail. However, if you can address them all, you will have an easier time with your project, and you will get more effective feedback before it is too late.

5.3 Writing scientific papers

One of your main tasks in this class is to learn how to write scientific papers. Being able to write well is a critical skill. You may not write all the time for your profession unless you are a journalist, a novelist, or a lawyer, but it is very likely that your future will depend in part upon your skill in writing. Most skilled jobs require frequent messages and memos. Any grants for which you want to apply, any promotions, and any awards, will depend upon written statements. Your career demands that you learn how to write.

Scientific writing provides good practice because what you wish to say is usually fairly straightforward, and you can focus on making your explanations clear, rather than searching for a story to tell. When most people think of scientific papers, the word "clear" is unfortunately probably not the first that comes to mind. The word "unintelligible" is probably more likely to come up. However, it is just wrong that scientists are deliberately obscure when they write about their work. They are trying to be as clear as possible, and the survival of their research demands that they explain it well. At the same time, every scientific field has a great deal of specialized vocabulary that scientists learn and use without thinking, but that can make it very difficult for a nonspecialist to follow the argument. This cannot be completely avoided, since special apparatus and situations demand special words, but scientists could probably use less specialized vocabulary than they do and improve the accessibility of their papers. At any rate, no matter what may be the strengths and weaknesses of typical scientists, your first task in writing a scientific paper is to be as clear as possible.

Some scientific journals demand that papers be written in a specific format. For example, *Proceedings of the National Academy of Sciences* requires papers to have a Title Page, Abstract, Introduction, Results, Discussion, Materials and Methods, Acknowledgments, References, in exactly that order. *Nature* does not allow an abstract, but the first paragraph essentially serves for one, and the journal editors require that the first paragraph have some sentences of very specific form, including one that begins "Here we show that. . . ."

The recommendation in this class is that when your paper describes an extended project, it have the sections of Title and Author, Abstract, Introduction, Procedure, Results, Conclusion, References. If you have a good reason to adopt a different format, consult with your instructors. Each of the sections is described below.

5.3.1 Title and author

The title of a paper should provide the best possible description of its contents in a few words. When scientists are searching through the scientific literature for information on some topic, the first thing they will see about a paper is its title. Based on the title, and maybe based on reading the abstract, they will decide whether it might be worth reading the whole paper.

It is very tempting to make the title of a paper something cute. In a class where the instructor has promised to read every paper anyway, this will not make a real difference, but in the scientific world, a cute uninformative title can put a barrier between the work and people who want to find it.

	Example	Comments
Don't	*Some Like it Hot*	This could be the title of a paper about tomato growth. You might want to check with your instructor before settling on a title like this. If your instructor has no sense of humor, it will not be allowed. If your instructor has a sense of humor, it might be allowed.
Do	*The Effect of Temperature on Growth of Tomatoes*	You might settle on this title after determining that your instructor has in fact no sense of humor.
Do	*Lichen as an Indicator Species for Air Pollution in Austin*	A good informative title from a student paper.

	Example	Comments
Do	*Dynamics of Static Friction between Steel and Silicon*	Title of Yang *et al.* (2008). The paper was submitted with the title *Dynamics of Static Friction*. The editors wrote back "We do, however, have a suggestion regarding your title. We thought *Dynamics of Static Friction between Steel and Silicon* would add some specificity and would be helpful to our diverse readership."

5.3.2 *Abstract*

Most scientific papers begin with an *abstract*. The goal of the abstract is to provide a compact summary of the main results. The main mistake you are likely to make in the abstract is to get it mixed up with a paragraph from your proposal. The abstract is not supposed to record what you thought or planned before you started your investigation, although it leads off your paper. It should probably be the very last thing you write. Imagine that someone will read only one or two paragraphs about your work, and you want them to learn as much as possible from a short description. Abstracts should be comprehensible when read without any additional information. They have no pictures or figures in them, and they rarely have citations

	Example	Comments
Don't	I will grow tomatoes at three different temperatures. My hypothesis is that tomatoes will grow the fastest at room temperature and will grow slower at higher temperatures.	This abstract is written as though the author has not yet performed the experiment. Perhaps it is taken from a proposal and has not been modified. The abstract describes a hypothesis that is not very specific and does not reflect any of the conclusions actually reached by doing the experiment.
Do	I have investigated the effect of three types of fertilizer on the growth rate of two types of beans. Each fertilizer was used at four different concentrations on four different samples of each type of bean. For each fertilizer, a concentration was determined that maximized the rate of plant growth. The optimum concentration of fertilizer was the same for the two types of bean.	This abstract is written in past tense, which is appropriate, since the experiment has been conducted and the answer is known. In passive voice the first sentence would become "The effect of three types of fertilizer on the growth rate of two types of beans has been investigated." Either form of the first sentence is acceptable.

	Example	Comments
Do	This paper presents empirical estimates of selection strength and migration load in a pair of three-spine stickleback populations that represent a remarkably close analog to a two-island model under migration selection balance. A larger deeper basin is connected to a smaller shallower basin by a short channel that allows extensive migration between populations. . . . Since the populations differ in size, theory would predict that the smaller population should experience stronger migration load. We confirmed this prediction . . .	From Bolnick *et al.* (2008). The language is a bit more technical than in a typical student paper, since the main audience is fellow scientists in the same field.
Do	We conducted experiments where steel and silicon or quartz are clamped together. Even with the smallest tangential forces we could apply, we always found reproducible sliding motions on the nanometer scale. . . . . The samples first slide and then lock up even when external forces hold steady. One might call the result "slip-stick" friction.	From the abstract of Yang *et al.* (2008).

of other articles. Particularly when writing abstracts, scientists avoid the first person, and use passive tense instead. Classes on writing usually recommend that one avoid the passive tense, but scientists use it anyway. Since the abstract describes what happened in your experiment, do not use future tense.

5.3.3 Introduction

Following the abstract, many papers begin with an introduction that lays out background and motivation for the work. This section of the paper usually explains why the work was done, and describes previous work that is relevant to the current investigation. There are usually many citations in this section. When citing prior papers, scientists lay out the main accomplishments in the previous published work, and then explain why because of (or despite) past accomplishments, new work was needed. These explanations can range from the observation that the previous work opened up new avenues for discovery, to the claim that the previous work was wrong and its errors need to be fixed.

	Example	Comments
Don't	Global warming is the largest threat faced by mankind. In a few years rising temperatures may make it impossible for people to feed themselves with the plants to which we are accustomed. In this experiment it was decided to investigate the effects of rising temperatures on tomato growth rates.	There are sweeping claims that are not backed up with any citations. Furthermore, the large claims are only loosely connected with the experiment.
Do	An example of human impact on the environment comes from lichen abundances and distributions around metropolitan areas. Lichens are particularly sensitive to air pollutants. L. Davies *et al.* (2007) found in their study of epiphytes (larger class of plants that includes lichens), that there was a marked decline in the diversity of lichens in the presences of high levels of NOx (oxides of nitrogen).	This student paper does a good job of spelling out technical terms needed to explain the background for the experiment.
Do	Species inhabit landscapes characterized by spatially variable climate, prey, and natural enemies. As a result, natural selection is also spatially variable, favouring different adaptations across a species' range (Hanks & Denno, 1994; Kawecki & Ebert, 2004). This tendency toward local adaptation is counteracted by gene flow.	From Bolnick *et al.* (2008). Sets up the context of the article, and backs up assertions with references to prior work.
Do	The understanding of friction has evolved greatly in the last 70 years [1,2.3]. Bowden and Tabor [4] established that friction is due to populations of asperities, and that the actual contact area of two solids in frictional sliding is much less than apparent. Dieterich [5,6,7] showed that the population of frictional contacts evolves during dynamic sliding, and measured changes in force over time, and at different steady speeds.	From Yang *et al.* (2008). As in many papers, this is a rushed attempt to summarize a large amount of prior work in a small amount of space.

5.3.4 Procedures

Your goal in describing the procedures you followed is to help the reader understand exactly what you did. To achieve the right level of detail, you should include enough information to make it possible for someone else to reproduce your experiment. When you are conducting the experiment you should make note of everything you can think of about it that might possibly be relevant, even if you are not sure at the time it is important. There are many experiments where the temperature of the room might turn out to be significant, for example. When you write up the paper, you should include those features of the experimental setup that you decided in the end were significant, and that someone else would need to make note of in repeating it. One or more pictures are often the best way to describe the way you set up an experiment.

For an extended example of describing experimental procedure, please see Appendix B, which contains an extract from Galileo's *Dialogs on Two New Sciences*. This book established some of the standards by which science has been

	Example	Comments
Don't	First you get your plants, and when they have grown at the temperatures, then you measure your data.	Far too much information is missing. The number and type of tomatoes is not described. The temperatures at which they were grown is not described. Precisely what was measured about them and when is not described.
Don't	Get a piece of wood. Cut it 2 inches wide. Cut it 4 feet long. Put a block 1 inch high under one end. Put a car at the high end.	The sentences are very short and all have exactly the same simple form, which makes the description unpleasant to read. In addition, many instructors will prefer that you use past tense to describe what you did, or present tense to describe what happens in experiments like yours, rather than giving the reader directions in the imperative tense.
Do	I used a clear acrylic rectangular box measuring $6 \times 6 \times 36$ inches as my water channel. The box had one open face, and holes at one end to allow water to drain out. The box was placed over a lab sink at a slight incline, with the draining end facing down. In addition, a feeder tube was attached to the sink faucet to channel water to the top of the box. A diagram is shown below (Figure 1): 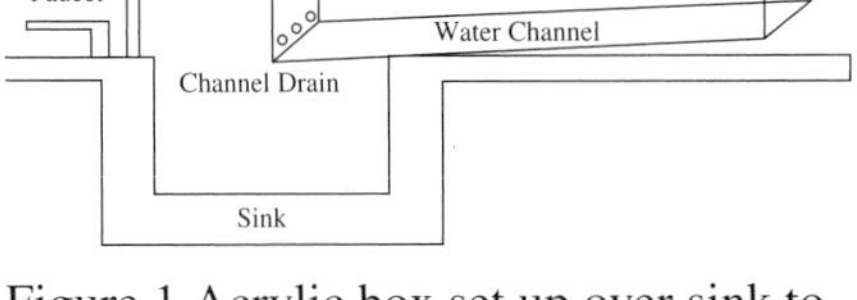	The written description together with the picture make the experiment clear. It would be even better to indicate distance measurements on the figure, and to specify the height of the "slight incline."

Figure 1 Acrylic box set up over sink to study liquid flow instabilities.

	Example	Comments
Do	We collected zooplankton in late morning of June 5, 2005 using three tows of a Wisconsin-style quantitative zooplankton net (100 μm mesh, 30 cm diameter) per basin. We immediately froze specimens on dry ice, and stored them at −80 °C.	From Bolnick *et al.* (2008). The language is very specific and contains many relevant experimental details.

conducted since, and provides a model of how to describe a scientific experiment so that it can be repeated. Use past tense to describe what happened, and present tense to describe what is always true. Avoid future tense.

5.3.5 *Analysis*

In this section, you describe the data you gathered and explain whether they are significant or not. To explore significance will usually require you to carry out statistical analysis, ranging from presentation of error bars on graphs to computations of t and χ^2 tests and the resulting p values. During the exploration of your own data you should find whether the statistical tests really make sense to you or not. It is important that the meaning of the tests be spelled out clearly in words. But remember that the statistical tests are not ends in themselves. They are tools you can use to support a story you want to tell about your experiment.

Statistical tests are not the only forms of analysis. Particularly in experiments concerning physics and chemistry, your experiment may be described by a precise mathematical theory. In this case, the analysis should contain a comparison of the theoretical results with what you have found. For example, suppose you are bouncing a ball and measuring the height after successive bounces. You might expect that the ball would lose some fraction f of its energy on each bounce, in which case the maximum height and potential energy of the ball at each bounce would be $1 - f$ times the maximum of the previous bounce. Thus in this case it would make sense to use the first few bounces to determine f and then check whether the successive bounces were in fact diminishing by a factor of $1 - f$ each time. In some cases you might be able to construct a mathematical model of the sort discussed in Section 4.6 and compare with data you have gathered. You should write a complete description of what you have done, and prepare graphs to illustrate the results.

In papers that have a theoretical focus, the analysis and procedures are not clearly separate. The goal in writing a theoretical paper, however, is not different from the goal in experimental papers. The goal is to present results in enough detail

	Example	Comments
Don't	I ran my t test and I got $p = 0.03294823$.	This might be acceptable if part of a larger discussion, but it is not acceptable by itself, mainly because there is no mention of the significance of the test. The fact that far too many digits of p are quoted lends to the suspicion that the writer views it as a meaningless number.
Don't	No statistics were possible in this experiment because too little data was gathered. However I am confident that the conclusions are valid, and flowers with MiracleGro really are larger.	If you have not gathered enough data to use any statistical tests, you should not be turning around and boldly claiming you found something significant anyway.
Do	I compared the size of lichen 1.5 miles from the center of pollution in Austin with those at the center. The lichen in the center had a mean size of 1.5 $\pm$.08 mm, while those 1.5 miles away had a mean size of 2.3$\pm$.2 mm. A t test showed that these means were significantly different ($p = 0.004$).	Good example of student work in which the difference between measured quantities is backed up with a statistical test.
Do	Stickleback density was estimated to be 74% higher in Dugout basin than Ormond basin (198 versus 82 fish caught after 98.5 and 72 trap-hours, respectively, yielding 2.01 and 1.15 fish per trap-hour, $\chi^2 = 18.26$, $P = 0.0001$).	From Bolnick *et al.* (2008). Here analysis indicating statistical significance of findings is woven into the description of the data. The Discussion section of this paper focuses on larger ideas and implications of the data rather than dwelling on the statistical significance of individual findings.
Do	Experiments by Dieterich [3,4,5] led in the 1970s to a new picture of friction, called *rate and state friction*. Two surfaces in contact are described not only by their relative velocity v but also by a state variable θ that evolves as the surfaces slide. An expression for the ratio of horizontal force F to normal force N for a sliding object is $$\frac{F}{N} = A \ln\left(\frac{v}{v^*} + 1\right) + B \ln\left(\frac{\theta}{\theta^*} + 1\right).$$	Opening of the analysis from Yang *et al.* (2008). The discussion lays out theoretical equations against which some experiments will be compared.

that an interested reader can reproduce them. It is very important to explain all the assumptions behind the calculations. It is almost always appropriate to display final results in graphical form, even if the final result is an equation.

5.3.6 Conclusions

The concluding section offers you the chance to summarize what you have found and make any final comments. Frequently, authors take the opportunity in the concluding words to mention the significance of the work, even if all the rest of the paper has focused on details of the project.

	Example	Comments
Don't	Despite my data said that rising the temperature did not affect the growth rate of tomatoes, or made it larger, I am sure that in reality tomatoes would be damaged. Global warming is a huge problem, every citizen needs to do their part to prevent it.	The researcher is discounting data in favor of expectations. He jumps without justification to strong conclusions about a large problem on which this particular experiment has little direct bearing. There are also several grammatical errors.
Do	An interesting follow-up study would be to investigate the exact levels of NOx, and SOx to the South and North as compared to the East and West to see if there is a difference that corresponds to the lichen abundances. As Austin grows and the number of commuters increases it will be important to take into account the effects pollution will have on our surrounding environment. When planning public transportation systems or maintaining city park quality, lichen might be the yellow canary in the mine.	Speculating on follow-up studies is fine. Curiously, thoughts about follow-up are very common in student papers, although the students are unlikely ever to follow up, and rare in published scientific papers, although the scientists are very likely to follow up.
Do	In conclusion, we suggest that morphological divergence between basins in Ormond Lake is constrained by gene flow, resulting in persistent maladaptation. Migration–selection balance can be viewed as a game of tug-of-war on an adaptive landscape: populations in different environments reside on distinct fitness peaks, and migration is the rope with which they pull each other closer together, and off their peaks.	From Bolnick *et al.* (2008). The authors close with a metaphor to make clear the basic idea of the paper.

	Example	Comments
Do	We expect our results to be significant for micro-machines that involve frictional contact between some of their parts, and whose tolerances require control below the submicrometer scale, for on these scales static friction as normally understood does not always exist.	From Yang *et al.* (2008). The very end of a paper can be a place for somewhat speculative comments about the significance of the work.

5.4 Scientific figures

Figures provide the most effective way to convey large amounts of scientific information quickly. They are also called graphs or plots, and spreadsheet programs refer to them as charts. The word "figure" usually refers to a polished graph, ready for publication. Sketches of scientific results prepared during the investigation are called "graphs" or "plots," and the word "chart" only appears in spreadsheets. Figures are so central to the presentation of research that some scientists write a paper by preparing the figures and then writing the text. There are many different types of figures, and a full discussion of how to prepare them can be quite long (Tufte, 2001). The graphs you are likely to prepare for this class fall into just a few categories and the following discussion aims to describe how you should think about preparing them. Chapters 2 and 3 are filled with examples, and the examples here are added to warn you away from common errors.

5.4.1 Choice of axes

Your graphs should have two axes. The horizontal or *x*-axis describes an independent or control variable, while the vertical or *y*-axis describes a dependent or response variable. You should think about cause and effect in settling on your axes. The horizontal axis should be something you have controlled, or something whose effect you want to measure when it changes. So, if you measure the time a ball needs to drop versus distance, you should put time on the vertical axis and distance on the horizontal axis, because in each experiment you controlled the distance and measured the time.

The quantities you plot on the vertical axis will always be numbers. Making observations quantitative is one of the central ways that science sets itself apart from other sorts of investigations. However, the quantities on your horizontal axis do not have to be numbers. For example, you might wish to measure the

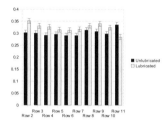

Figure 5.1 **An example of what not to do!** This plot contains many errors. It is a plot of the data from Table 2.2. The control variable is presence or absence of lubrication, so there should be two categories on the horizontal axis. Instead all the individual trials are plotted, and error bars are placed on each trial. The fonts are too small, there are no axis labels, and there is no caption or figure number.

heights of students depending on whether they are men or women. The students are divided into two *categories*, men and women, and you represent the results with a *categorical plot*. Categorical plots can have many more than two categories. If you measure a functional relationship, such as the time it takes a ball to fall versus distance, then the quantities on the horizontal axis are numbers that represent the continuously varying control parameter. This is a *continuous plot*. For a falling ball, the horizontal axis would contain numbers describing various distances.

5.4.2 Categorical plots

A common mistake in the preparation of categorical plots is to put individual measurements rather than the categories that interest you on the horizontal axis. An example of this mistake is shown in Figure 5.1. The figure shows data from Table 2.2 on the time needed for a ball to roll down a channel depending on whether or not the channel is lubricated. The categories are "Unlubricated" and "Lubricated" but Figure 5.1 makes the mistake of plotting every single trial. If you plot the data in this way, you are delivering the incorrect information that the first unlubricated and first lubricated trial need to be compared with each other, and are somehow essentially different from all the other trials. The main reason you would prepare a plot like this is that spreadsheet programs create them very easily. If you highlight columns of data and click on a menu item to create a chart, Figure 5.1 is what you are likely to get. Don't do it.

While you are carrying out an experiment, you should regularly prepare scatter plots such as Figure 5.2. The horizontal axis shows the control variable in which you are interested, and the vertical axis shows the response variable. One

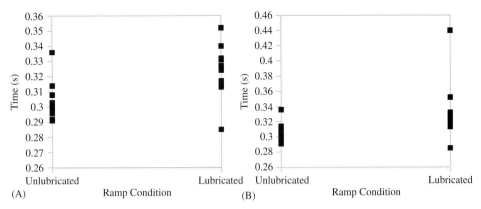

Figure 5.2 Scatter plot. (A) During the course of an experiment, you should regularly prepare scatter plots. The horizontal axis shows the important categories for your experiment, and the vertical axis shows the quantity you are measuring. Each individual experimental trial is represented by a separate point. (B) Catching outliers. A data point whose value is 0.34 has been mistyped as 0.44, and stands out from others in a scatter plot.

reason to prepare scatter plots is that they let you quickly catch experimental points that do not fit a pattern. Suppose that when typing in the data for experiment, in one case you type 0.44 rather than 0.34. The scatter plot in Figure 5.2(B) makes this point leap out from the others. You can double check your work and decide whether the point is legitimate or not. Points that stand out like this and break a pattern are called *outliers.* A common question when someone sees an outlier is whether it is alright to discard the point. Usually, the answer is "no." For example, if you give a test where the mean is 70 and most students have scores between 60 and 80, you cannot throw away the score of a single student who gets 100 on the grounds it is an outlier. There should be some reason to discard the point other than the fact that it is unexpected. You should be able to point to part of the procedure that broke down, some step that you forgot or carried out incorrectly. It is legitimate to repeat the measurements that led to the outlier, in an attempt to decide whether it really occurred, or resulted from experimental errors.

Scatter plots sometimes end up in publications, but they have the defect of providing a little too much information for rapid understanding. Most commonly, researchers prepare plots that focus on the mean and standard error, as in Figure 5.3. You may feel a bit cheated by this plot, because it does not show all the work you did by performing many different trials. The point of a scientific publication is not to explain to everyone how much work the scientist did, but to convey the results. This is the most compact way.

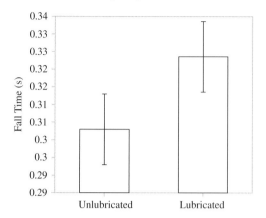

Figure 5.3 The data of Table 2.2 can be conveyed rather well by typing two numbers and their error ranges. However, if one wants to plot them, the plot should
look like this, and not like Figure 5.1. This sort of categorical plot is very common
in publications. See Appendix A for guidance on creating such plots in spreadsheet programs, and particularly on how to add customized error bars.

5.4.3 Continuous plots

Continuous plots represent functional relations. The horizontal axis represents a
numerical quantity that can vary continuously, and the vertical axis represents
a function of it. The goal of the plot is give a visual sense of the functional
relation.

One choice you must make in preparing such plots is how or whether to connect
the points together with lines. There are basically three choices:

1. Plot just points, not lines. You may choose to avoid connecting points with any lines
   in scatter plots, where the points are scattered and you are not sure what functional
   relation they are describing, as in Figure 5.4.
2. Pass a trend line through the points. An example showing how to do this appeared as
   Figure 4.5. It is also possible to completely miss a pattern hiding in data by proceeding
   in this way. For example, suppose the data of Figure 4.7 were to be plotted as a scatter
   plot with a regression line. The result is shown as Figure 5.5.
3. Connect all points to their neighbors, either with straight line segments or smoothed
   line segments. All the plots in Figure 4.3 are of this type (like all plots prepared by
   computer, they are based on computations at a finite number of points) as are Figures
   4.7 and 4.10. This type of plot can bring out patterns in data. But it can also obscure
   them. If the data in Figure 4.5 are replotted by connecting all points with lines, the
   result is an ugly function where the rapid jumps up and down have no significance
   (Figure 5.6).

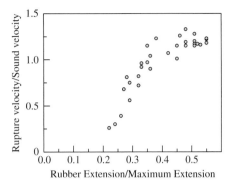

Figure 5.4 Continuous scatter plot. Speed of running ruptures in rubber versus how hard the rubber was stretched. There is clearly a functional relationship, but the data are noisy, the function does not look familiar, so the points are not connected with lines.

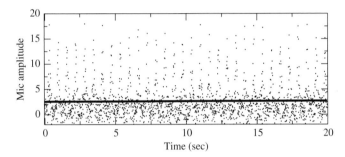

Figure 5.5 An example of what **not** to do. Data of Figure 4.7 replotted as a scatter plot with trend line. All information about the oscillations in the data is lost.

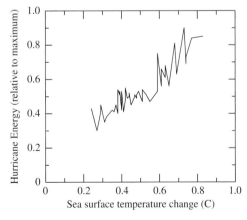

Figure 5.6 An example of what **not** to do. Data of Figure 4.5 replotted by connecting all adjacent points with lines.

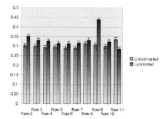

Figure 5.7 **A final example of what not to do!** Data from Table 2.2 are plotted again making many unfortunate decisions. Colors (red and blue) were used to distinguish the two values of the control parameter, but unfortunately the paper was run through a copy machine and the color was lost. Matters are made even worse by keeping the gray background and gridlines that are the default in charts from spreadsheet programs. The color gradients added to each bar also do not help.

5.4.4 Other ingredients of figures

Someone reading a scientific paper may just be flipping through the paper looking at the figures. Therefore, the figure must be accompanied by a caption that makes the figure as clear and self-contained as possible. On the other hand, they may be reading through the text of the paper. Therefore, the figure needs to be numbered, should be referred to specifically in the text by number, and should appear in the text close to the first reference.

It is tempting to use color in order to distinguish different elements of your data. However, be warned that some readers may only see your paper after it has been copied, in which case all the color will be lost. Then they will face making sense of a figure like Figure 5.7. Some final tips appear in Table 5.1.

5.5 Giving scientific presentations

The three main ways that scientists communicate their work are through presentations (also called talks, seminars, colloquia, or lectures), articles, and books. Most scientists have reasonably heavy travel schedules and give on the order of five to fifty presentations per year. These range from short contributed talks at conferences (2–10 minutes) to invited talks at conferences (20–60 minutes), to seminars, colloquia, and public lectures at research institutions (usually 60 minutes). Seminars are intended for specialists in a field and are highly technical, colloquia are supposed to be aimed at an audience including undergraduates majors, while public lectures are supposed to be accessible to a broad literate public. No matter what audience scientists speak to, they tend to make their talks too technical. After someone has been working on a problem for years, it is hard to remember all the different points

Table 5.1 Guidelines for figures in a scientific paper.

Do	Don't
Label axes.	Use color unless you are sure that all readers of the figure will see the colors.
Make fonts large enough to read.	Accept default choices from spreadsheet programs.
Number figure.	Create graphs quickly using chart wizards in spreadsheet programs.
Write a reasonably self-contained caption.	Leave graphs embedded in spreadsheets and refer to them in the text of your paper.
Think about cause and effect before settling on x and y axes.	Paste a figure from a spreadsheet into the text without a caption and without checking that fonts and graphics are of the right size after pasting.
Prepare scatter plots during the experiment even if you eventually use another sort of figure in the final paper.	Wait until the night before a project is due to plot data for the first time.
Try several sorts of plots to represent data in order to find which explains them best.	Submit the first plot you prepare whether or not it represents your work well.
Put categories of interest on the horizontal axis.	Represent each trial individually on the horizontal axis.
Plot means and standard errors when appropriate.	Plot a point for every trial on the vertical axis if means and standard errors are appropriate.

that seemed confusing at first, became obvious later, and explain them well in a short time.

In this class you will give a number of short presentations of your work, probably no more than 3–8 minutes in length. You will not be describing a project you have pursued for years, but you will have been working on it for weeks, and therefore you will face the challenge every scientist experiences of helping an audience understand quickly what you accomplished. Since your fellow students have probably been working on very different projects from yours, you should prepare your talk as a public lecture, assuming that few in the audience have much prior knowledge about the subject you are presenting.

The format of scientific presentations has changed rapidly in the last decade. Most talks are accompanied by computer-projected slides using PowerPoint, Acrobat Reader (for pdf documents), or similar programs. Some talks, particularly in mathematics, are still given at the chalkboard. You will probably use PowerPoint to give your presentation, as most scientists do.

Graphs and sketches are even more important in presentations than in papers. Sketches provide the best way to explain your experimental setup and graphs provide the best way to convey your data. A suggested outline for your talk is

Introduction and motivation: Explain what motivated you to carry out your project.
Background: Provide information that typical listeners will need in order to understand what you did.
Design and procedures: Explain what you built or used in order to investigate your question.
Results and analysis: Explain the data or results you obtained, and whether or not they are significant.
Conclusion: Provide final comments you would like to make, including thanks and acknowledgments.

In a very short talk, such as one that lasts two minutes, you may be able to afford only a sentence or two in order to touch on these elements; there may not be time for much background, but you should give thought to each of them as you prepare. Table 5.2 provides hints on what to do and not to do in a presentation.

5.6 Searching for scientific information

For beginning scientists, the idea of publishing a scientific paper seems remote and daunting. Finishing problem sets is hard enough. What could it possibly be like to do scientific research, something original, and actually have it published in a real scientific journal?

Writing a scientific paper that truly changes the way that people view the world, founds a new discipline, or gives birth to new industries, is truly difficult. But simply performing a piece of publishable research, and getting it into print, is not terribly different from the inquiries and papers you are completing in this class. Your inquiries and write-ups have all the ingredients of the professional counterpart: twists and turns, a constant feeling of uncertainty, the gap between original intentions and final accomplishments, simple errors that come back to haunt you after you have solved ten times more difficult problems along the way, the important result that comes out easily, almost by accident and is nearly overlooked, the impossibility of something that seems easy, the trivial solution of something that seems hard, and the terribly unexpected reactions of the outside world to one's results.

Really, the only difference between the inquiries in this class and publishable research is that for your inquiries the instructors may largely be indifferent to

Table 5.2 Guidelines for scientific presentations.

Do	Don't
Prepare slides that consist mainly of pictures with a few words for guidance.	Paste a figure from a spreadsheet or paper into the presentation without checking that fonts and graphics are of the right size after pasting.
Focus slides on main points you would like to make.	Paste tables full of numbers into the presentation.
Reformat plots prepared for a paper with larger fonts so that numbers and axis labels are easy to see across the room.	Type long stretches of text into the presentation.
Explain ideas behind computations and analysis.	Display long equations.
Prepare multiple slides rather than putting too much information on individual slides.	Use colored backgrounds that make it difficult to read text or see graphs.
Speak to the audience rather than looking at screen.	Avoid eye contact with the audience.
Slow down and speak loudly.	Speak very quickly and softly.
Practice the presentation with friends beforehand and accept suggestions for improvement.	Give the presentation for first time ever to a live audience.
Time the presentation to make sure it fits into the time allotted.	Prepare slides without regard to the time available.
Arrive early to make sure that the presentation is compatible with presentation software and hardware in room.	Show up at the last minute and hope for the best.

whether the results are original or not, and the time-frame for each inquiry is limited to a few weeks. In the great world of science, the results of an inquiry should actually be new, and the time-frame is a little longer, around one to five years, which is roughly the time that companies, funding agencies, or universities will grant before demanding to see results.

This means that scientists have to be able to figure out what has been done before, so that they can spend their time building on previous work, rather than repeating it. Nothing is more deadly for an attempted publication than for a reviewer to point out that key results have already been published.

This seems like an easy matter to deal with. A scientist just needs to stay current with articles in the field where he or she is working. The problem is that the number of articles is so large, and the places where they could appear so varied, that staying current becomes more difficult all the time. Here is an example to show how publication in science has been increasing.

The rectification of current – that is, the fact that electrical current sometimes flows more easily in one direction than another – was discovered in 1874 by Schuster and Braun. Writing in 1947 on the subject of metal rectifiers, Henisch (1957) wrote that after 80 years of work on the problem

The student of rectifier problems is confronted with an extensive literature, including some 80 to 100 papers of major historical, practical, or theoretical importance.

Now, jump ahead to 1986, when Bednorz and Müller (1986) discovered high-temperature superconductivity. By 1988, the literature on high-temperature super-conductors already included 5200 articles, by the year 2001, over 73,900 articles on the subject were in print, and by 2008 the number increased to over 98,000. No human being can possibly read all these articles or know in detail what each of them contains.

So your goal will almost certainly not be to find all existing articles on a topic. However, you will be carrying out assignments that require you to find some articles on a topic, preferably a few that are clear, authoritative, give an overview of a field, and answer some questions you have posed. Back in olden times 10 years ago, this required spending days or weeks in the library, or finding an expert in the subject who could give tips. Today it means carrying out computer searches.

Most universities make powerful specialized search engines available through their library websites. A brief set of these is mentioned in Table 5.3. For most purposes, if it is available, ISI Web of Knowledge should be your first choice. It has broad coverage of all of science and mathematics back to around 1975. One of its most appealing features is the ability to search for articles citing ones you have found. So, if you find an article that seems useful, you can find the articles it builds upon by looking at the references, and you can find the articles that build upon it by looking at those that cite it. This search feature enables you to examine the progress of science backward and forward in time.

No two university library websites are identical, but the following description of how to find articles on high-temperature superconductivity at The University of Texas at Austin is probably very similar to what you will need to do:

1. Go to `www.lib.utexas.edu`.
2. Under `Research Tools`, click on `Databases and Indexes to Articles`.
3. Under `Databases Alphabetical by Title`, click on `I`
4. Click on `ISI Web of Science (ISI Web of Knowledge)`
5. In the top search box, type `"high temperature" supercond*`

Table 5.3 Brief list of selected online databases of scientific publications.

Subject	Name	Comments
All of science	ISI Web of Knowledge	Very broad coverage, makes it possible to follow citations.
All of science	Google Scholar	Freely available at scholar.google.com. Makes it possible to follow citations.
Astronomy	INSPEC	Broad coverage of physical sciences. Indexing goes back to nineteenth century.
Biology	PubMed	PubMed is freely available to the public at pubmed.org.
Chemistry	Chemical Abstracts	Can search by chemical compound.
Earth Sciences	GeoRef	
Engineering	Ei Compendex	Searches technical reports and conference proceedings other services miss.
Mathematics	MathSciNet	On-line access to Mathematical Reviews database, can follow citations.
Medicine	PubMed	PubMed is freely available to the public at pubmed.org.
Physics	INSPEC	Broad coverage of physical sciences. Indexing goes back to nineteenth century.

The result in 2010 is around 22,000 articles, less than the nearly 100,000 located by the more specialized database INSPEC, but quite enough for most purposes. Putting `high` and `temperature` next to each other in quotation marks means that these two words must be found next to each other in that order in a description of the article. The asterisk (`*`) after `supercond*` is a *wild card*. It matches both `superconductor` and `superconductivity` for example; anything beginning with `supercond`. These conventions for searching are common, but not universal. Other databases may assume that words next to each other must be found in that order unless you put AND in between them. Whenever you use a new database you should make sure to learn its conventions for searching.

When carrying out literature searches, you will constantly need to zoom in and out until the scale is right. Make the search too general, as in `high temperature superconduct*`, and so many articles pop up that the result is overwhelming. Make it too specific, as in `high temperature and superconduct* power cable Texas`, and no hits come up at all. A query at the right scale gets a few hundred results, as in `high temperature and superconduct* power cable`, which pulls up 146 articles describing the current effort to make power transmission lines out of high-temperature superconducting materials. Most of the recent articles are by Japanese and European authors, which is one reason why restricting the search to Texas is a mistake.

If you do not have access to services such as ISI Web of Knowledge and want something freely available, there are two options. A large portion of the medical and biological literature (with smatterings of other topics) can be searched at `pubmed.org`. Google seems to hope to replace all these services with Google Scholar, at `scholar.google.com`. Google Scholar is particularly good at digging up freely available preprints on the web, and like Web of Knowledge makes it possible to follow citations. It is not as comprehensive as Web of Knowledge, but has the advantage of being free to the world.

5.7 Obtaining scientific articles

The distribution of scientific information has undergone a rapid transition from printed volumes stored on library stacks to electronic media that scientists access over the internet. New issues of every serious scientific journal are now distributed electronically, and some journals have made all their back issues available electronically as well. Once you have found the title of an article that interests you and want to obtain the whole thing, you will be faced with a number of possibilities:

- The article may be just one click away. The search engine you have used to find the article may provide a link to the full text, and if the article is free, or if your university has paid for access, you may be able to obtain it immediately.
- If you cannot obtain full text in this simple fashion, you may still have access through your university. The journal may be accessible in electronic form through a different page. Note the journal name, volume, and page number of the article you wish to find, locate the list of electronic journals your university makes available, and search there. For example, at the University of Texas, go to `www.lib.utexas.edu`, under `Research Tools` click on `Find a Journal` and then type the name of the journal. Even if you obtain information from the university library that the issue you are looking for is not available, keep moving forward since journals are being digitized so quickly library information is not always current.
- Check Google for the journal. Some journals are not available though library websites because they are free to everyone.
- If the work involves biology or medicine it may have been funded by the National Institutes of Health (NIH). In this case the article must be made freely available through PubMed (Figure 5.8).
- Check Google Scholar for the article title and author names. Many authors submit a pre-publication copy of their work to electronic preprint servers such as `arXiv.org`,

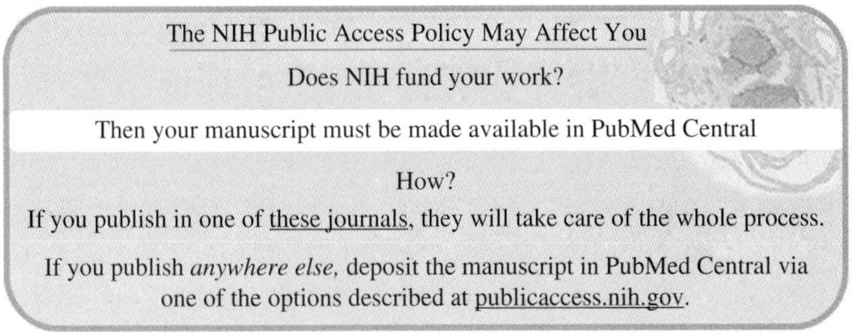

> The NIH Public Access Policy May Affect You
>
> Does NIH fund your work?
>
> Then your manuscript must be made available in PubMed Central
>
> How?
> If you publish in one of <u>these journals</u>, they will take care of the whole process.
>
> If you publish *anywhere else,* deposit the manuscript in PubMed Central via
> one of the options described at <u>publicaccess.nih.gov</u>.

Figure 5.8 A sign of the times. Information for authors on the home page of `pubmed.org`.

and you can find something very close to the published article that way. This is most common in physics and less so in other disciplines.

- If all else fails, find a library, go to it, get the article, and read it. If you have never wandered the library stacks of journals in your discipline, you should do it at least once. Nothing else gives a sense of how large the research effort is in science. Every page of every article in the endless stacks of huge volumes represents weeks, months, or years of work by teams of people, and for most journals has been read by at least one and often more than one reviewer.

5.8 Reading scientific papers

Once you have actually obtained scientific papers (also called articles) on a subject that interests you, you will face two final problems. Many articles are very hard to read. And they do not all agree with one another. So the two problems are making sense of the papers, and deciding which to trust.

In deciding what to trust, you should give very reduced weight to any article that has not been peer reviewed. The peer review process is essentially identical to the process you have experienced in partner grading of your inquiries. When the editor of a journal receives an article, he or she sends it to one or more experts in the field for evaluation. Within a time-frame of about a month, the editor asks for a written report on the article, describing its strengths and weaknesses, asking for improvements, and asking whether on balance it should be published at all. Two referees submitted careful reports on Yang *et al.* (2008), which included statements such as

How were s_f and s independently measured? This is not clear enough in the paper...

So if equation 1b were a generalization of Dieterich's theory, then I would have expected something of the following kind: $dT/dt = (1 - T/T_0) - Tv/D_c$ but NOT equation 1b, ... If this is really what was used, then I fear that I cannot recommend publication...

Since the abstract is crucial I take the liberty to make some suggestions...

The manuscript was accepted for publication after additional experiments were performed in response to referee questions and each of their questions and criticisms was addressed in detail. The peer review process is not always this thorough. Sometimes reviewers read articles quickly late at night and fail to catch mistakes. They have to take on faith some claims of authors, since it is impractical for reviewers to repeat all experiments and all calculations in papers they receive. Serious scientific mistakes have survived peer review and been published (Levi, 2002). Still, the process is the best there is. Peers review papers to see if they are right, and you should not trust papers, particularly on controversial topics, that have not made it through peer review.

How can you tell if a journal is peer reviewed? The standard reference work is Ulrich's Periodical Directory, `http://www.ulrichsweb.com/`. Accessing information from this source requires a subscription; many university libraries subscribe, so you may have access if you are a student or researcher at a university. Typing in the name of a journal as a quick search on the upper right side of the page, you will get back summary information about the journal, including whether or not it is peer reviewed.

Among peer reviewed journals, not all have the same authority and influence. The most prestigious journals are *Science* and *Nature*. Each of these journals covers all of science, although they emphasize the biological sciences. They contain both short articles, called letters, and a small number of longer research articles, as well as news stories about science, reviews, and other features. Within every discipline there is a variety of journals. Some publish only short articles on important rapidly moving topics. Others publish only long articles that contain thoughtful overviews of a field. For example, in physics, *Physical Review Letters* publishes short articles on rapidly moving topics, *The Physical Review* publishes longer articles to give all the details of a completed research project, and *Reviews of Modern Physics* publishes long summaries of the results of many research articles. The broadest readership however belongs to popular journals such as *Scientific American*, *Discover*, or *Sky and Telescope*, which draw upon experts but are not peer reviewed. A short list of leading journals in various disciplines appears in Table 5.4.

How can you make sense of articles? Popular journals are the easiest to read, but also the least authoritative. It sometimes seems that the more comprehensible is the account of a subject, the less one is allowed to trust it. This is not completely accurate, since articles that have great influence in science have to be well written enough for other scientists to understand them. But it is a real problem nevertheless. *Science* and *Nature* have come up with a partial solution. Accompanying most of the letters they publish are shorter pieces, written by other scientists, that summarize the results, explain them, and put them in context. The combination of an original article together with its explanation can be a very effective

Table 5.4 A very small selection of journals of different types for mathematics and science.

Discipline	Rapid publication, peer reviewed	Archival, peer reviewed	Semi-popular, partly peer reviewed
All sciences	*Science, Nature*	*Science, Nature*	*Science, Nature*
Astronomy	*ApJ Letters, Astronomy Letters*	*Astrophysical Journal, Astronomy and Astrophysics, Monthly Notices of the Royal Astronomical Society*	*Physics Today, Spark: The AAS Education Newsletter*
Biology	*Science, Nature*	*Proceedings of the National Academy of Sciences USA (PNAS), Cell, Journal of Biological Chemistry*	*ASBMB Today*
Chemistry	*Chemical Physics Letters, Nano Letters*	*Journal of the American Chemical Society, Chemical Reviews, Surface Science Reports*	*Chemical and Engineering News*
Mathematics	*Mathematical Research Letters*	*Annals of Mathematics, Bulletin of the American Mathematical Society, Acta Mathematica*	*Mathematical Intelligencer*
Physics	*Physical Review Letters, Europhysics Letters, Applied Physics Letters*	*Physical Review, Reviews of Modern Physics, Journal of Applied Physics*	*Physics Today*

way to make sense of a scientific topic. In the case of review articles the author is charged with trying to explain and summarize a large number of more technical articles. The summaries are often easier to follow than the articles being summarized. Finally, many scientists prepare websites on which they explain their work in clear language. You will have to search for these with Google since the scientific articles themselves never mention them.

Sometimes the only barrier to reading a scientific article is a fairly small number of technical words. Once you have looked these up, the article is no longer so hard to follow. At other times an article may presume that you know material that could only be learned in graduate-level coursework, and in this case the article is bound to be hard going. You will have to use your judgment to decide whether a particular article is worth the investment of time needed to read it.

Here are some additional suggestions for approaching technical articles.

- Expect to spend more time and effort reading a scientific article than an article in the newspaper.
- Read the title and abstract carefully, and if there are hints that the article is interesting, look up all words you do not know. If the authors have done their job properly, the essence of the article is conveyed in the abstract.
- Read the introductory section, paying particular attention to the articles or reference works the authors cite as introductions to the field. Even if their own article is on a specialized topic that is not essential for you, the references may point you to clear overviews of the field that are more helpful.
- Look at the figures or data tables and read the captions carefully. Often the main results of an article can be learned this way; the text explains how they were obtained.
- When you run across symbols that look unfamiliar, keep in mind the possibility that it is unfamiliar notation for an idea you understand. In a well-written paper, the text should describe the symbol well enough that you will be able to figure this out.
- Many articles are written by people for whom English is not a native language, and in some cases the quality of the language can be very poor. If you see so many grammatical errors that you cannot understand what the authors are saying, it is probably not worth your while to continue.

5.8.1 Controversial topics

When investigating a controversial topic in science, there is a resource you should know about. The National Research Council has the task of supplying scientific information to Congress and every year convenes panels of scientists to provide expert advice on topics ranging from the medical uses of marijuana to the best methods of preparing science teachers. After a period of months to years, the panel conclusions are released as books through the National Academy Press. Nearly 2000 books are available at www.nap.edu, covering almost every scientific question that affects public policy. Ask whether power lines cause cancer, what is the best food for horses, and whether creationism is science, and you will find answers from the scientific establishment.

For example, searching on "Creationism" gives

1. Science, Evolution, and Creationism (2008, 88 pp.). Committee on Revising Science and Creationism: A View from the National Academy of Sciences.
2. In the Light of Evolution: Volume 1. Adaptation and Complex Design (2007, 380 pp.). John C. Avise and Francisco J. Ayala, Editors.
3. Evolution in Hawaii: A Supplement to Teaching About Evolution and the Nature of Science (2007, 56 pp.). Steve Olson, The National Academies.
4. Quantum Leaps in the Wrong Direction: Where Real Science Ends...and Pseudo-science Begins (2001, 200 pp.). Charles M. Wynn and Arthur W. Wiggins, With cartoons by Sidney Harris.

5. Science and Creationism: A View from the National Academy of Sciences, Second Edition (1999, 28 pp.). Committee on Science and Creationism, National Academy of Sciences.

6. Responsible Science, Volume I: Ensuring the Integrity of the Research Process (1992, 224 pp.). Panel on Scientific Responsibility and the Conduct of Research, National Academy of Sciences, National Academy of Engineering, Institute of Medicine.

7. Teaching About Evolution and the Nature of Science (1998, 150 pp.). Working Group on Teaching Evolution, National Academy of Sciences.

8. Fulfilling the Promise: Biology Education in the Nation's Schools (1990, 168 pp.). Committee on High-School Biology Education, National Research Council.

9. High-School Biology Today and Tomorrow (1989, 364 pp.). Committee on High-School Biology Education, National Research Council.

Science is international, and for some scientific topics the best representation of the scientific consensus may come from international organizations. For example, for questions of global warming, the dominant international voice is the Intergovernmental Panel on Climate Change, at `www.ipcc.ch`. Numerous summary reports can be obtained from this website.

5.9 Final words

At its heart, science is simple. Science is about checking. Many people believe that opinions on hosts of subjects are personal choices, and that there is nothing more to say. "People born in the winter are really more inquisitive than everyone else, but that's just my opinion." Science listens to such statements and replies, "All right, let's check." Find a sample of people born in the winter. Define *inquisitive*. Find a way to measure inquisitiveness. Check.

Problems arise along the way. There is always some uncertainty. The samples are not as large as one might like. There are variations from one measurement to another. Scientists have developed experimental procedures and mathematical techniques to tease as much certain knowledge out of uncertainty as possible.

The statistics, mathematical models, conventions on drawing figures and writing papers are not ends in themselves. They serve the goal of checking things people want to know and explaining what was found. Some things people want to know just because they are curious, others in order to solve problems or manufacture products. The core of scientific knowledge contains assertions that are thousands of years old and very unlikely ever to change. The periphery of scientific knowledge contains controversies, uncertain statements, arguments that switch back and forth in months or days.

When you learn scientific research methods, you are learning that not everything is just a matter of opinion. You are learning how to develop the most certain knowledge of natural phenomena that humans have ever obtained. Whether you aim to be a teacher, a scientific researcher, or a citizen setting out in other directions, this knowledge should change how you look at the world.

Assignments

5.1 Inquiry

Purpose Carry out a final project, incorporating skills gained during class.

Background You have carried out a number of inquiries of different types during the semester. Your goal is now to combine the skills gained in these different inquiries into a final project. These projects begin with *curiosity*, proceed with *experimental design and taking of data*, continue with *statistical analysis* and *modeling*, and the ability to *access the scientific literature*. You now should combine what you have been learning in a final project.

Length 5–10 typed pages. It is generally not a problem to exceed this length.

Report Please include the following sections

    a. Title.
    b. Abstract.
    c. Introduction.
    d. Experimental design.
    e. Analysis, including an appropriate statistical treatment of the data.
    f. Conclusions.
    g. References. You should make use of the peer-reviewed literature as appropriate. Unless your instructor indicates otherwise, you should cite at least three peer-reviewed sources.
    h. Data either in the body of the report, or as an appendix.

Grading This inquiry will be evaluated according to the criteria in the Inquiry Grading Rubric, Appendix D.

5.2 Literature search 1

In this assignment, you will practice finding articles in the scientific literature. Suppose you want to find information on global warming. A direct search on the words "global warming" uncovers such a large number of articles that it is also valuable to restrict the search also to subtopics. Here are examples of more specific questions.

a. What are anticipated consequences of global warning for the Greenland ice sheet?
b. How far back in the past do ice cores provide information on average annual global temperatures?
c. How quickly, if at all, is the global sea level rising?
d. Are the habitats of butterflies changing measurably due to global warming?
e. How quickly are atmospheric levels of carbon dioxide changing?
f. How much is global temperature estimated to rise over the next 50 years?
g. What are possible consequences for human agriculture of a mean global temperature rise of $2\,°C$?

h. What are the primary "drivers" of global temperature change?

i. To what extent is global warming observed from satellites an artifact due to construction of paved urban areas?

j. What are global climate models, and are they reliable?

k. What international protocol to limit global warming is currently being negotiated? Should the United States participate and sign it?

l. If a frog is put into a slowly heated pot of water, does it jump out or does it let itself boil to death?

m. What is the cost per kilowatt hour of wind, solar, and wave power in the U.S., and what are environmental concerns generated by these technologies?

n. What is the cost per kilowatt hour of nuclear power in the U.S., and what are the safety concerns surrounding nuclear waste disposal?

o. How much radiation would be released in a nuclear power plant containment breach, and how many people would be impacted?

Choose one of these questions, or investigate other sets of questions if directed by your instructor. First, search using Google and describe the best information you find. Next, search using professional databases made available through your library and describe the best information you find.

You must turn in a brief description of the sorts of information you found in your Google and database searches, and a minimum of two complete references to recent articles from the peer-reviewed literature. For each article, locate the full text electronically, print out the first page, and hand it in.

5.3 Open questions literature search

In this assignment you will search for peer-reviewed literature on open and challenging questions in science or mathematics.

a. How close has a computer come to passing the Turing test, and what are the prospects of a computer passing the test in our lifetime?

b. Are embryonic stem cells necessary for scientific research?

c. How are planets detected around distant stars, and is it likely that any detected so far house life?

d. Why is the Big Bang theory of the universe viewed as more plausible than the Steady State theory?

e. What are the principles of quantum computing, and what are the main technical challenges still to overcome?

f. Describe some proposals for perpetual motion machines and explain whether or not such a machine is possible.

g. How is the human genome mapped, and what exactly does one learn by mapping it?

h. What is the evidence behind the claim that 70% of the mass of the universe is unknown, and what experiments are being done to track down the missing mass?

i. Is it possible or likely that avian flu will become a disease that passes from human to human?

   j. What are the main health risks caused by cell phones?

   k. How was Fermat's Last Theorem proved?

   l. What is the current status of the Goldbach conjecture?

  m. Is Kepler's conjecture about the most efficient way to pack spheres proved or not?

Choose one of these questions, one of the questions in Problem 5.2, or investigate other sets of questions as approved by your instructor. First, search using Google and describe the best information you find. Next, search using professional databases made available through your library and describe the best information you find.

You must turn in a brief description of the sorts of information you found in your Google and database searches, and a minimum of two complete references to recent articles from the peer-reviewed literature. For each article, locate the full text electronically, print out the first page, and hand it in.

5.4 **Debate**

  Background Your instructors will assign you to study a debatable topic involving science. You must prepare for a debate, making use of library sources, and paying attention to evidence that backs up opinions you express.

  Teams For each topic, there will be two teams. However, each team will learn only on the day of the debate which side of the debate it will be arguing. Therefore, you must research both sides of the topic to which you have been assigned.

  Format Sample format of the debate, depending on number of students and length of class period: to open, two minutes for each member of the PRO team and two minutes for each member of the CON team. Then, each member of the PRO team will have up to two additional minutes for rebuttal, and each member of the CON team will have the same. Finally, there will be five minutes for the audience to ask questions of any side it wishes.

  Written assignment On the day you conduct your debate you must turn in a list of sources you have consulted to prepare. You will only receive full credit for the debate participation if your sources include at least three articles for each side of the debate from the *peer-reviewed research literature*, and if you refer knowledgeably to articles as relevant in your oral presentation. For purposes of the written homework, turn in

    a. A total of at least six citations of peer-reviewed articles (title, author, journal, volume, page, year).

    b. For each article, at least one sentence describing a specific debating point the article supports.

  Topics Your instructors will assign you a debate topic to investigate. Within the context of the general topic of global warming, some specific possible topics include:

a. Mean global temperature and mean global sea level are rising at a rate that will pose dangers to human societies within the next 100 years.

b. Human carbon dioxide emissions are largely responsible for global warming, and must be curtailed.

c. The United States should rapidly expand its use of nuclear power so as to reduce the emissions of greenhouse gases from oil-, gas-, and coal-fired power plants.

5.5 **Open questions paper**

Write a 2–3 page paper on an open or challenging question in science or mathematics. You may adopt a question from the lists provided in Assignments 5.2 or 5.3, or with instructor approval choose another question of your own. You must make use of a minimum of four articles from the peer-reviewed literature. The paper must cite the articles and provide clear evidence that you have read the articles and make use of them in formulating your arguments.

5.6 **Open questions presentation**

In this assignment, you will give an oral presentation concerning an open or challenging question in science or mathematics. You may adopt a question from the lists provided in Assignments 5.2 or 5.3, or with instructor approval choose another question of your own. Consult your instructor to learn the time available for the presentation. Make sure not only to try to answer the question, but also to ask why it is important, who is interested in it, and why.

References

G. Bednorz and K. A. Müller (1986), Possible high T_c superconductivity in the Ba-Ca-Cu-O system, *Zeitschrift für Physik*, **B64**, 189.

D. Bolnick, E. J. Caldera, and B. Matthews (2008), Evidence for asymmetric migration load in a pair of ecologically divergent stickleback populations, *Biological Journal of the Linnean Society*, **94**, 273. Copyright 2008 Wiley Blackwell, extracts reprinted with permission.

H. K. Henisch (1957), *Rectifying Semi-Conductor Contacts*, Clarendon Press, Oxford.

B. G. Levi (2002), Investigation finds that one Lucent physicist engaged in scientific misconduct, http://scitation.aip.org/journals/doc/phtoad-ft/vol_55/iss_11/15_1.shtml, *Physics Today*, p. 15.

E. R. Tufte (2001), *The Visual Display of Quantitative Information*, 2nd edn, Graphics Press, Cheshire, CT.

I. Valiela (2001), *Doing Science: Design, Analysis, and Communication of Scientific Research*, Oxford University Press, Oxford.

Z. Yang, H. P. Zhang, and M. Marder (2008), Dynamics of static friction between steel and silicon, *PNAS*, **105**, 13,264. Copyright 2008 National Academy of Sciences, U. S. A., extracts reprinted with permission.

Appendix A

Spreadsheets for basic scientific computation

A.1 Spreadsheet programs

Spreadsheets record and manipulate data. The purpose of this appendix is to emphasize the features of spreadsheets that make them useful for manipulating scientific data, and to perform computations based on scientific theories.

A.1.1 Excel for Windows

By far the most common spreadsheet is Excel, which is developed and marketed by the Microsoft Corporation as part of Microsoft Office. There are many different versions of Excel in circulation, but all of them have all the functions described here. Excel is available for all the different Windows operating systems Microsoft has written for IBM-compatible personal computers. The major releases of Excel in current use are Excel 2003, and Excel 2007. The latest release, Excel 2007, involves a completely new way of accessing menu items, but all functions of Excel 2003 are still available. Note that the native file format of Excel 2007 is a new format (.xlsx) that can cause difficulties for other programs. Although this format is becoming universal, you may still sometimes find it useful to read and write files in the old Excel 2003 format (Compatibility Mode).

A.1.2 Excel for Macintosh

Excel 2008 is available for computers running Apple's OSX operating system. The menu items are arranged somewhat differently from the Windows releases, but all the same functions are available in a nearly identical way. This version of Excel can read the .xlsx files of the recent Excel releases for Windows.

A.1.3 OpenOffice.org

Excel is not available for Linux or other open source operating systems. However, there are open source programs that provide most of the same functionality as Excel, and can legally be installed for free on Linux, Windows, and Macintosh operating systems. The most widespread is the spreadsheet program in OpenOffice.org. The OpenOffice.org spreadsheet program can open most Excel 2003 files, and can store most spreadsheets in this format. OpenOffice.org does not provide all the features discussed in this appendix. None of the features of Excel's Analysis Tookpack are available, including its *t* tests and histograms. Nevertheless, if Excel is not available, OpenOffice.org provides an acceptable substitute. Note that Excel cannot read files stored in the native file format of OpenOffice.org (.ods); the Excel 2003 format (.xls) is the format to employ so that the maximum number of people can read and edit your spreadsheet. OpenOffice.org version 3 can read some .xlsx files, but cannot write in this format.

The figures used to illustrate spreadsheet operation here come at different times from Excel 2003, Excel 2007, and OpenOffice.org, to emphasize the essential similarities of all these programs. They do differ in many details, such as the precise location of menu items and how they interact with mouse clicks and keyboard shortcuts. You will have to experiment with these features. in order to make the most of the spreadsheet programs available to you.

A.2 Basic program features

A.2.1 Workbooks, sheets, and cells

Figure A.1 is a screen shot of an Excel 2007 workbook. A *workbook* is like a legal notepad containing many different sheets. When you open up a spreadsheet program, it opens up a blank workbook for you.

A *sheet* is a collection of cells arranged in rows and columns. If you look at Figure A.1, you will see that you are looking at Book 1 (look at the top of the window for the book label), and at Sheet 1 (look at the tabs at the bottom left of the workbook)

Data are stored in *cells*. Each cell is named by its physical position in the spreadsheet. Notice that Column A is highlighted as well as Row 9. Columns are vertical groupings of cells. All columns have a letter name that increases from left to right starting with A. Rows are horizontal groupings of cells. All rows have a number name starting with 1 and increasing from the top down.

If you want to talk about a particular cell then you have to use its name. The cell enclosed by a black border in Figure A.1 is cell A9. In general, the first part of the cell's name is the column that it is in and the second part of its name is the row that

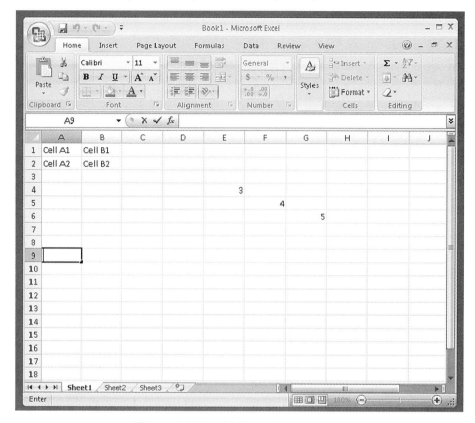

Figure A.1 Excel 2007 under Windows.

it is in. The names of cells A1 through B2 are indicated within them. What is the value of the data in cell G6 in Figure A.1?

A.2.2 Selecting cells

At any given time, some cells in your spreadsheet are highlighted, have been selected, and are active and others are not. You will need to select single cells in order to change the data in them. You will need to select multiple cells in order to copy from them, create charts, and operate with functions.

To select a single cell, left click on it with the mouse. You can also navigate between cells by using arrow keys.

You can select a column of cells by clicking on its letter. For example, to select column G, click on the gray G at the top of the window. To select a row, click on its number; to select row 19, click on the gray 19 on the far left of the screen.

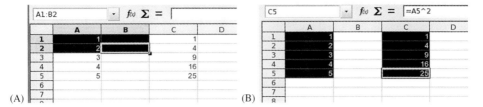

Figure A.2 (A) Selecting cells A1:B2. (B) Selecting cells A1:A5 and C1:C5.

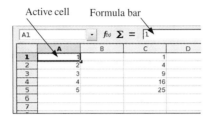

Figure A.3 Active A1 cell and the formula bar showing the data "1".

To select a collection of cells that are next to each other, click and drag with the mouse. For example, to select cells A1, B1, A2, and B2, click the mouse in cell A1 and drag it to B2 (Figure A.2A). The name for this collection of cells is A1:B2. Whenever you see two cells separated by a colon, the first cell gives one corner of a rectangular collection of cells, and the second one gives an opposite corner of the rectangle of cells.

You can select cells that are not next to each other by holding down the control key (Apple key, for Macintosh) as you click and drag with the mouse. Try to select A1:A5 and C1:C5 simultaneously, as in Figure A.2(B).

A.2.3 Changing data

Once you have selected a single cell with a single click, anything you type will replace the current contents of the cell. The contents of the cell also show up inside the *formula bar,* shown in Figure A.3 . If you want to edit the contents of the cell rather than just replacing them, you can click repeatedly on the cell until a cursor becomes active inside and you can move back and forth with arrow keys typing and deleting. Or, you can click inside the formula bar and edit the contents of the cell there.

A.2.4 Adding and deleting rows and columns

You may find that you want to insert data between cells that are already filled in. To do this you need to insert rows or columns. You can do this by selecting the

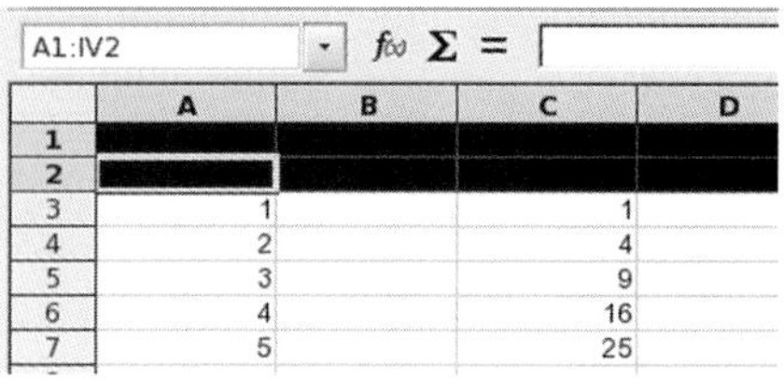

Figure A.4 To insert two rows, select two rows (A) and then right click and choose `Insert` or `Insert Rows`, depending on your version of the spreadsheet.

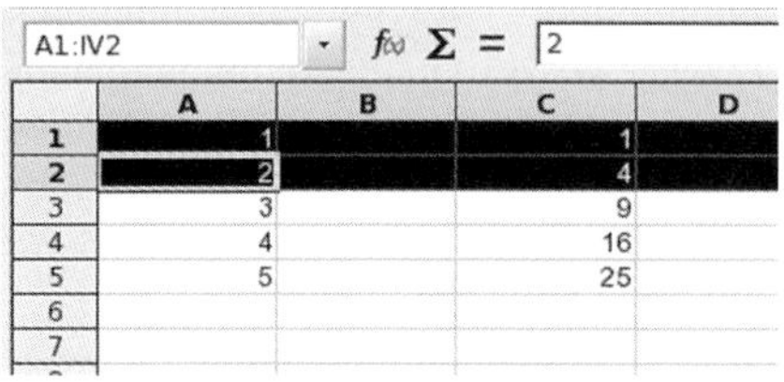

number of rows you wish to add, right clicking, and choosing an `Insert` menu item as shown in Figure A.4. A similar procedure applies to columns. Sometimes you want to delete a column or a row. To do this you can select the row or column and press the Delete key, or right click, and choose a menu item indicating deletion.

A.2.5 Rapid navigation and selection

If you deal with large spreadsheets, you may need to move up and down sheets with tens of thousands of rows. Pressing the `Ctrl` key and then ↓ , the DOWN ARROW key will take you immediately to the last non-blank entry of the current column, while `Ctrl` and then ↑ will move you to the top. `Ctrl` ← and `Ctrl` → similarly move to the beginning and end of the current row. By pressing `Shift` and `Ctrl` plus arrow keys simultaneously, it is possible to select whole rows and columns rapidly.

A.2.6 Formulas

Spreadsheets can perform calculations for you. To tell them to do this you need to enter a formula into the cell for which you want the value to be displayed. To tell Excel to start computing a formula, the information in a cell must begin with an = sign. In Figure A.5(A), cell B5 does not exactly have the number 25 in it; instead, it has the formula =A5^2, which means "Square the contents of cell A5 and display the result."

A.2.7 Copying formulas

One of the most powerful features of a spreadsheet is connected with the way it reacts when you copy the formula in one cell and paste it into another cell. After pasting, all the cells that take part in the formula are shifted by the distance between the original cell and the one to which you have pasted it. For example,

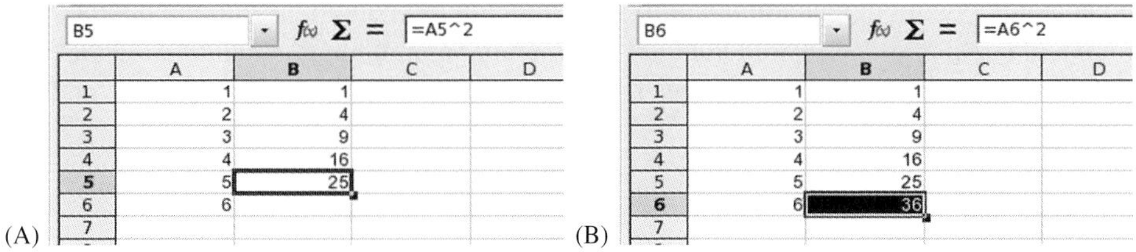

Figure A.5 Formula =A5^2 is copied into cell B6 and changes to =A6^2 automatically, as shown in formula bar. To perform this operation, select cell B5 by clicking on it and copy the contents of the cell, either through the menu item Edit->Copy, or through the keyboard shortcut Ctrl-c (Apple-c, Macintosh). Next select cell B6 by clicking on it and paste the formula into it, either through the menu item Edit->Paste, or through the keyboard shortcut Ctrl-v (Apple-v, Macintosh). The formula automatically adjusts to refer to the next cell down.

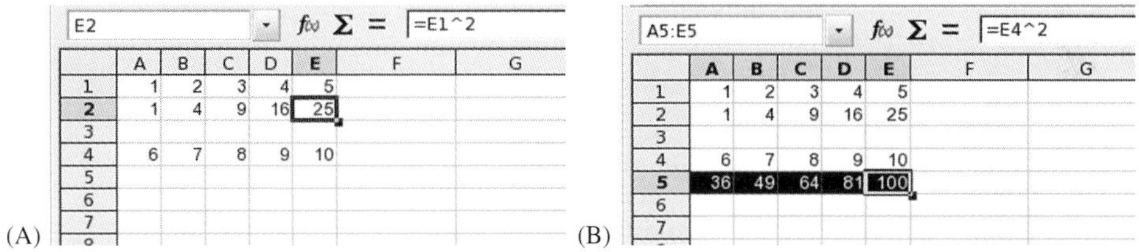

Figure A.6 Copying the formula from cell E2 into cells A5:E5 produces formulas for the squares of all the integers in the cells above. In (A), select cell E2 and copy it. In (B) the contents of E2 have been copied into A5:E5.

in Figure A.5, you paste cell B5 into cell B6. The formula in cell B5 is =A5^2. But when you paste it into cell B6, it becomes =A6^2, because it is shifted down by 1. The same sort of logic applies when you paste into cells that are shifted both vertically and horizontally. In Figure A.6, you copy from cell E2 into cells A5:E5. The formula =E1^2 becomes =A4^2 after pasting into A5, =B4^2 after pasting into B5, and so on. In each case, the formula refers to the cell that is just one cell above the one into which you are pasting. You can see the cells to which a formula refers by making a cell with a formula active, and clicking inside the formula bar; the spreadsheet will highlight all the cells that are part of the formula.

Spreadsheets' ability to modify formulas when copying them from cell to cell gives them the ability to implement recursion relations of the sort studied in Section 4.6.

A.2.8 Static references to cells

Sometimes you will want to create a formula that keeps the row or the column or both constant when you paste it into a new cell. This is called a *static reference*. You can create one by putting a dollar sign $ in front of the row or column name

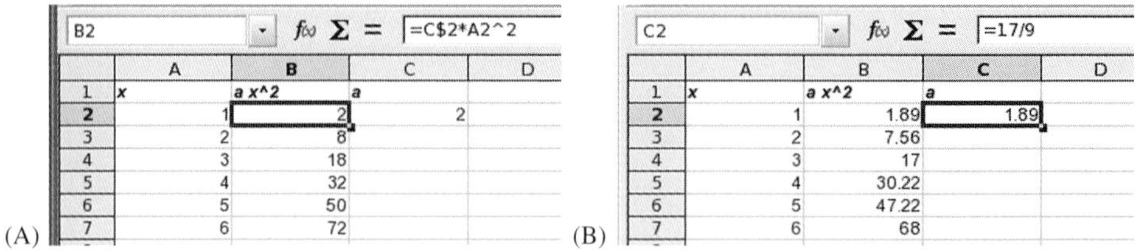

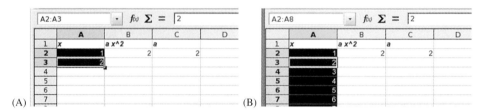

Figure A.7 (A) Formula with a reference to cell C$2. The dollar sign means that
when the formula is copied into cells B3, B4, and so on, the row "2" does not
change. Therefore, column B implements the formula ax^2 where x is in column
A, and the constant a is in cell C2. The advantage of implementing the formula
this way is shown in (B). If cell C2 is changed to 17/9 (which is computed as
a formula), all the cells in column B are immediately updated to reflect the new
value of the constant a.

Figure A.8 (A) Highlight two cells with the numbers 1 and 2 in them. There is
a small black square at the lower right-hand side of the highlighted region. Place
the mouse pointer over it; the cursor should change to a cross. Click and drag
downwards over as many additional cells you want to fill in. The spreadsheet
figures out that you probably want to create a series of increasing numbers, and
fills in the new cells as shown in (B). When cells contain formulas, spreadsheets
paste in formulas just like copy and paste commands.

that you do not want to change. For example, suppose you want to compute the
formula ax^2 for a number of different values x, but also to be able to vary the
constant a. Figure A.7 shows how to accomplish this. When you paste from cell
B2 into cells B3:B7, the formula continues to use C2, because by writing it as
C$2, the row number is not allowed to change. If you wanted to make sure that
neither the row nor the column could change no matter where you pasted your
formula, you would enter the cell as $C$2.

A.2.9 Fill command

Spreadsheets offer a fast way to copy values from one cell to another with the
mouse, and try to figure out your intentions as they goes along. Differing spreadsheet programs may make somewhat different guesses, or prompt you for your
intentions. Figure A.8 shows how this is done.

A.2.10 Paste special

Sometimes you want to copy the numerical value of a cell, but not the formula. In order to do this you need to make use of the "Paste Special." In Excel 2007 or 2008, it is found by selecting Home on the ribbon, and clicking on Paste. In OpenOffice.org, or older versions of Excel, choose Edit → Paste Special. To copy and paste the numerical value but not the formula, first select the cells that you want to copy, then copy them. Now select the area in which you want to paste the values and select Paste Special. Then click on the Values button and finally click OK or hit Enter.

One occasion when Paste Special is useful arises when you create a list of random numbers and want the list to stay the same. If you do not capture the numbers with Paste Special, they will be updated and changed any time you change any feature of the spreadsheet.

A.3 Built-in functions

Spreadsheets have many built-in functions. To browse through the list of functions click on the Formulas tab in Excel 2007, 2008 and select the leftmost option on the ribbon, or Insert → Function in OpenOffice.org. A menu will then pop up that lists all the categories of functions that you can use. Most used in this course are Statistical. After selecting a function click OK. A window will pop up guiding you to provide arguments.

An example of the built-in AVERAGE function was shown in Figure 2.2. The function wizard from OpenOffice.org that guides you in using this function appears in Figure A.9. You do not have to use the function wizards, but they make it much easier to remember the arguments needed by the built-in functions than simply typing them in.

Data analysis toolpack Excel has some supplementary functions in the Data Analysis Toolpack. These come for free with Excel, but you must install them. For Excel 2007, click on the Office Button 🔘, select Excel Options, and look for Add-Ins. For older versions of Excel, on the Tools menu, select Add-Ins. Next select Data Analysis, and say OK. You may be prompted for the Office installation disk. OpenOffice.org does not have these functions.

Multiplication When you multiply two numbers together, you must always use an asterisk * between them. If you type =5 2 you will get an error. You must type =5*2 to get 10.

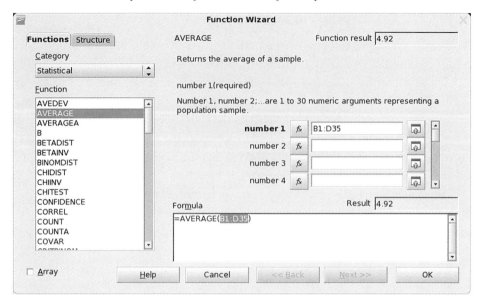

Figure A.9 The function wizards from Excel or OpenOffice.org can guide you through the process of averaging a group of numbers. You can select which numbers to average by clicking inside the box labeled **number 1** and then using the mouse to highlight the desired cells within the spreadsheet. Clicking on the arrow just to the right of this entry box reduces the size of the function wizard box and makes it easier to select cells. Click on the arrow a second time to make the wizard come back. When you are finished, click OK, and the formula =AVERAGE(B1:D35) will be entered in the cell you have selected, which in the case of Figure 2.2 is cell B38.

Addition, subtraction, division These are given by the ordinary symbols $+$, $-$, and /, as in =5/2 which returns 2.5.

Order of operations Spreadsheets perform multiplication and division first, and addition and subtraction afterwards. So =5/2+3 returns 5.5, not 1. When in doubt, use parenthesis. For example =5/(2+3) returns 1.

Precision Spreadsheets carry out operations with 15 places of precision, but only display as many as requested, with the default being two past the decimal point. You can change the number of digits displayed by formatting the cell, for example with the menu option Format → Cell.

Powers and square root To raise any number to any power, type for example =5^(1/3), which raises 5 to the 1/3 power. You can also use =SQRT(3) to find the square root of three.

Sum The sum of numbers in some range such as B1:D35 is =SUM(B1:D35).

Count The number of numbers in some range such as B1:D35 is =COUNT(B1:D35).

Nesting You can put functions inside functions inside functions as much as you like. For example, you can compute =5/3*SQRT(AVERAGE(B1:D35)) to find five-thirds of the square root of the average of the values in B1:D35.

Average The average of numbers in some range such as B1:D35 is =AVERAGE(B1:D35).

Standard deviation The sample standard deviation of numbers in some range such as B1:D35 is =STDEV(B1:D35).

Standard error The standard error of numbers in some range such as B1:D35 is =STDEV(B1:D35)/SQRT(COUNT(B1:D35)).

Z tests Excel provides the built-in function ZTEST. It takes three arguments. The first is a range of data, such as A1:J10. The second is the value against which the mean of the data is to be compared. The third is the standard deviation of the distribution with which the data are to be compared. Prior to Excel 2007, the documentation for ZTEST claimed that it returned the p-value for a two-tailed Z test when in fact it returns the p-value for a one-tailed test. Figure A.10 shows how to compute the p-value for a two-tailed Z test from the function ZTEST.

	A15				f_x	=2*ABS(0.5-ZTEST(A1:J10,0.5,0.5))						
	A	B	C	D	E	F	G	H	I	J	K	L	M
1	0	0	1	0	0	0	1	0	1	1			
2	0	0	0	1	1	1	0	0	0	1			
3	1	0	1	1	0	0	1	0	1	1			
4	1	1	1	0	0	1	1	1	0	0			
5	0	0	1	1	0	0	1	1	0	1			
6	0	0	0	0	0	0	1	1	0	0			
7	0	0	0	1	1	0	1	0	1	0			
8	1	0	1	0	0	1	1	1	0	1			
9	1	1	0	0	0	0	0	1	1	1			
10	0	0	0	0	0	1	1	1	1	0			
11													
12													
13	0.46												
14	0.501												
15	0.58												
16													

Figure A.10 Two-tailed Z test applied in Excel 2007 to the coin flip data of Table 2.4.

A.3.1 *t tests*

Spreadsheets provide several separate ways to compute probabilities from *t* tests. There are two built-in functions, TDIST and TTEST. The first of them takes three arguments; the value of *t*, the number of degrees of freedom *d*, and whether to perform a one-tailed or two-tailed test. For the comparison of a ball in lubricated an unlubricated channels in Table 2.2 on page 20, the function would be invoked as =TDIST(2.7,18,2) and it returns a value of 0.014. A separate function TTEST does almost all the work from scratch. It acts directly upon the array of experimental data, as shown in Figure A.11. The first two arguments are the groups of data to be compared. The third argument tells the function whether to compute a one-tailed (1) or two-tailed (2) test. The final argument chooses between three separate variants of the *t* test. The first variant (1) assumes that the data are paired. For example, if you measure acidity of water at 10 locations in a cave, clear brush away from the top, and measure water acidity at exactly the same locations afterwards, it would be appropriate to use a paired test to see whether the brush clearing had a significant effect. The second variant (2) assumes that the variances of the two distributions are the same; for example, in the ball drop experiment. The third variant (3) assumes that the variances of the two distributions may be different, as for the sizes of fish in Lakes Ormond and Dugout. The Data Analysis Toolpack contains three different variants of the *t* test report, corresponding to the assumptions that data are paired, variances are the same, and variances are different.

χ^2 *tests*   Spreadsheets provide two ways to carry out χ^2 tests. The first of them, CHIDIST is similar to ZDIST and TDIST. It takes two arguments, the first of which is the value of χ^2 and the second of which is the number of degrees of freedom *d*. It returns the probability of χ^2. For example, to compare the fitting function and experiment for the flight of the bottle rocket described in Table 3.8, compute =CHIDIST(6.25,7) to find the probability that chance alone could explain the differences.

The function CHITEST implements the form of the χ^2 test described in Section 3.8.2. It operates upon two tables of data containing observations and expectations, as shown in Figure A.12, and provides the probability of χ^2 as output.

Random number generator   Sometimes it is very useful to have spreadsheets simulate noisy data by generating random numbers. The function =RAND() will generate a random number from 0 to 1. To generate a random number from 0 to 100 simply multiply the random number by 100, =RAND()*100. To generate a random number between two numbers *a* and *b* type =RAND()*(b-a)+a. Note that spreadsheets will find new values for every random number anytime you do anything that

A2 | f_x | =TTEST(A2:A11,B2:B11,2,2)

	A	B	C	D	E	F	G
1	Unlubricated	Lubricated		t-Test: Two-Sample Assuming Equal Variances			
2	0.303	0.352					
3	0.301	0.331					
4	0.292	0.327					
5	0.296	0.315					
6	0.304	0.313					
7	0.291	0.317					
8	0.314	0.332					
9	0.308	0.34					
10	0.298	0.324					
11	0.336	0.285					
12							
13	0.014						
14							
15							

t-Test: Two-Sample Assuming Equal Variances

Input
Variable 1 Range: $A$2:$A$11
Variable 2 Range: $B$2:$B$11
Hypothesized Mean Difference:
☐ Labels
Alpha: 0.05

Output options
⦿ Output Range: $D$1
○ New Worksheet Ply:
○ New Workbook

OK Cancel Help

(A)

TTEST | X ✓ f_x | =TTEST(A2:A11,B2:B11,2,2)

	A	B	C	D	E	F	G
1	Unlubricated	Lubricated		t-Test: Two-Sample Assuming Equal Variances			
2	0.303	0.352					
3	0.301	0.331			Variable 1	Variable 2	
4	0.292	0.327		Mean	3.04E-01	3.24E-01	
5	0.296	0.315		Variance	1.74E-04	3.26E-04	
6	0.304	0.313		Observations	1.00E+01	1.00E+01	
7	0.291	0.317		Pooled Variance	2.50E-04		
8	0.314	0.332		Hypothesized Mean Difference	0.00E+00		
9	0.308	0.34		df	1.80E+01		
10	0.298	0.324		t Stat	-2.73E+00		
11	0.336	0.285		P(T<=t) one-tail	6.86E-03		
12				t Critical one-tail	1.73E+00		
13	=TTEST(A2:A1			P(T<=t) two-tail	1.37E-02		
14				t Critical two-tail	2.10E+00		
15							

(B)

Figure A.11 (A) Two ways to compute a t test. On the left is shown the function TTEST, which takes the first group of data A2:A11 as a first argument, the second group of data B2:B11 as a second, an argument that is 1 or 2 depending on whether one wants a one-tailed or two-tailed test, and finally an argument describing whether samples are paired (1), variances of the two distributions can be assumed the same (2) or variances of the two distributions may be different (3). The Excel Data Analysis Tool Pack can also produce a report including t tests. (B) shows the result of running the report, which contains the two sample means, t, and the probability p of t for both one-tailed and two-tailed tests.

changes a value anywhere in the spreadsheet. This can be very annoying. To get rid of this feature, once you have generated your list of random numbers simply copy your list of random numbers and paste them somewhere using the Paste Special command. Remember to select Values and not Functions.

B11		▾ ◎	f_x	=CHITEST(B2:E3,B8:E9)		
	A	B	C	D	E	F
1	Observed	Calc I	Calc II	Diff EQ	Lin Alg	
2	Engineering	30.00	45.00	75.00	150.00	300
3	Math	75.00	75.00	125.00	225.00	500
4						
5	Totals	105.00	120.00	200.00	375.00	800
6	Percentages	0.13	0.15	0.25	0.47	
7						
8	Expected	39.38	45.00	75.00	140.63	
9		65.63	45.00	125.00	234.38	
10						
11	p(chi^2)	0.21				
12						
13						

Figure A.12 The function CHITEST takes two arguments. The first is the matrix of observations, B2:E3, and the second is the matrix of expectations B8:E9. These data come from the sample survey on page 102. The result in cell B11 is the probability of χ^2. Note the warnings about using this form of the χ^2 test on page 101.

Histogram   Excel's Data Analysis Toolpack makes histograms (Section 3.2.3). It is up to you to prepare a column containing the bins in which you would like the data to be sorted, as in Figure A.13(A). To obtain data for this histogram, select the ribbon items Data → Data Analysis → Histogram. Figure A.13(A) shows a window that asks for the data you would like sorted, and the bins into which you want them to be put. Figure A.13(B) shows the result of clicking on OK and creating data for the histogram.

A.4 Charts

One of the powerful features of spreadsheets is their ability to make charts from data. Figure A.14 shows how to create a plot of continuous data, and Figures A.15 and A.16 show how to add error bars to each point.

A.4.1 Continuous data – XY scatter chart

Although spreadsheets offer the option of creating line charts, you will rarely use them. When you have calculated or measured a continuous function and wish to display it, you should use an XY scatter chart as shown in Figure A.14.

A.4.2 Adding custom error bars

All spreadsheet programs can add error bars of different sizes to each data point. Error bars can be added equally well to scatter plots, or to bars and columns

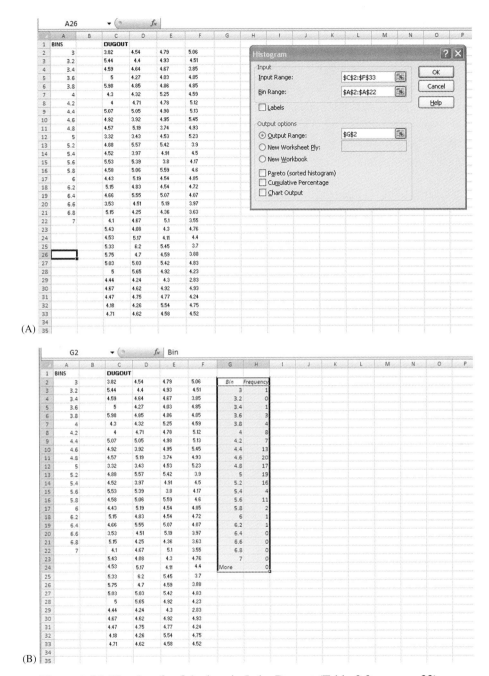

Figure A.13 The data for fish sizes in Lake Dugout (Table 2.3 on page 22) are used to demonstrate how to create a histogram with the Data Analysis Tool Pack. (A) shows the window that accepts information about data and bins, while (B) shows the output of the operation.

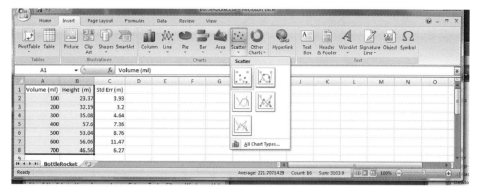

Figure A.14 To begin making a chart of continuous data, select the columns you wish to plot. These data show height of a bottle rocket versus the amount of water in it, and come from Table 3.8 on page 99. You can either click and drag the mouse to highlight exactly the data you wish to plot, or else just click on the column headers. For Excel 2007, click on the `Insert` tab and select `Scatter`.

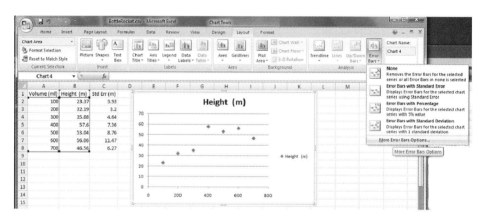

Figure A.15 To add error bars in Excel 2007 for Windows, click on `Layout`, and select `Error Bars`. Do **not** choose `Error Bars with Standard Error`, but instead choose `Custom`. In Office 2008 for Macintosh, double click on a data point to bring up the `Format Data Series Palette`, and then select `Error Bars` at the left. In OpenOffice, double click on your chart to select it, then right click on a data point obtain a context menu, select `Y Error Bars`, and then `Cell Range`. Excel 2003 for Windows is similar to OpenOffice.

in categorical plots. Although the programs provide an option to put error bars with standard errors onto your points, you should never use it, because what the programs compute and draw does not correspond to what scientists use. Instead, you must compute the standard error for every data point, and use it as shown in Figures A.15–A.17.

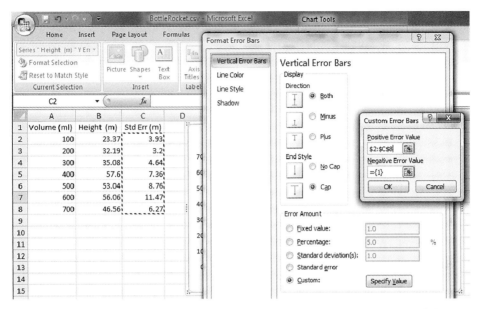

Figure A.16 To provide the heights of the error bars, select the cells containing values you have computed for the standard error of each point.

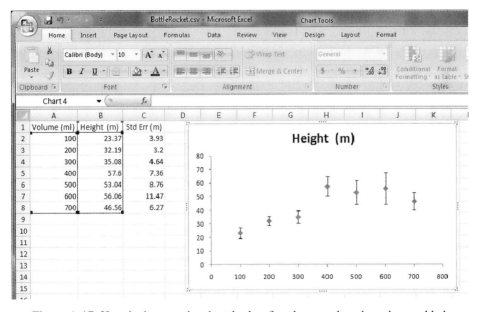

Figure A.17 Here is the way the chart looks after the error bars have been added.

A.4.3 Trend line

Spreadsheets can add trend lines to data sets, selecting from a menu of functions. This is one case where allowing a spreadsheet to do too much of the thinking seems not to be very productive. It is better if you come up with your own function that you wish to compare with data, prepare a column containing the function, and plot the two together yourself.

A.4.4 Categorical data – Column chart

Figures A.18–A.20 show some of the steps needed to create a chart from categorical data.

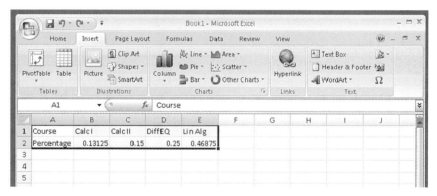

Figure A.18 To begin creating a chart from categorical data, select the data you wish to plot. This example shows how to create a column chart from percentages of students with different favorite math courses, Table 3.10. The first row will be interpreted as column labels.

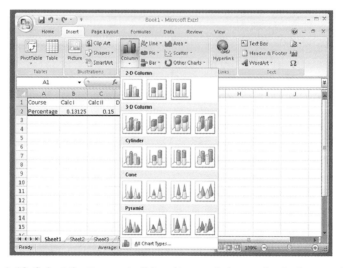

Figure A.19 Select the `Insert` ribbon item and choose the `Column chart`.

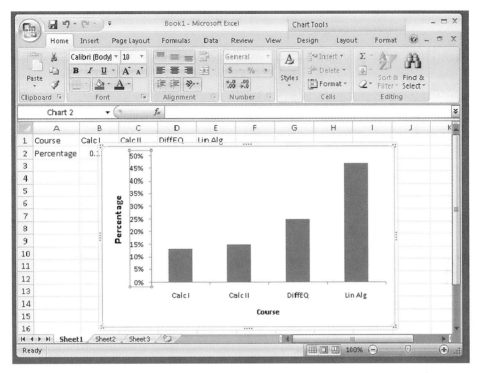

Figure A.20 After adding axis labels, deleting grid lines, and other formatting changes the chart looks as above.

Assignments

A.1 **Using functions**

  a. Input the data 1,2,3,4,5 in cells A1 through A5. Calculate the average using the average function and insert the result into cell B1.

  b. Calculate the sum using the sum function and insert it into cell B2.

  c. Calculate the standard deviation using the STDEV function and insert it into cell B4.

  d. Use a function to add 3.14 to each number in cells A1 through A5.

  e. Insert your results in cells C1 through C5.

  f. Copy all of your calculations and data and paste them (by formula, not by value) into cell A7.

  g. Format all of your data so that it displays only two decimal places.

A.2 **Data and graphing**

  a. Label cell A1 Person, label cell B1 Test1, label cell C1 Test2, label cell D1 Average, label cell E1 Standard Error.

  b. Enter the number 1 into Cell A2 and the number 2 into cell A3. Use the fill command to fill the column, through A6, with the numbers 3, 4, and 5.

  c. Now enter random test scores for each person in cells B2 through B6. Let the scores lie between 60 and 100.

   d. Do the same for cells `C2` through `C6`. (If you are adventurous, you will use the random number generator to do this for you.)

   e. Label cell `D1` `Average Score`. Use the average function to calculate the average test scores for each person and insert them into cells `D2` through `D6`.

   f. Next, find the standard error (for example for `E2`, calculate `=STDEV(B2:C2)/sqrt(2.0)` of each person's test scores, and place the result in cells `E2-E6`.

   g. Create a graph of average test score vs. person, using a scatter plot with points (no lines connecting the points). Please format and label the graph.

   h. Place individually sized error bars on each point, as described in Figures A.15, A.16, and A.17.

A.3 **Formulas** In this exercise, you will use a spreadsheet to look at the doubling problem. You are going to receive an allowance for 30 days. You have to decide how to take it. If you choose to have a Normal Allowance, you will get 10 dollars a day. If you choose to have a Doubling Allowance you get 1 penny the first day, 2 pennies the next day, 4 pennies the third day, and so on for a total of 30 days. Each day your allowance is double the allowance of the previous day.

   a. Label column `A` `Day`.

   b. Use the fill command to create 30 rows, with numbers 1 through 30 in rows `A2` through `A31`.

   c. Label column `B` `Normal Allowance`.

   d. In column `B`, create a formula in Excel that will calculate the total amount of allowance that you earn on day 1, day 2, day 3, etc. You can use the fill command to fill the numbers in rapidly. Note: you are earning $10.00 every day.

   e. In cell `A32` write `Total` and in cell `B32` enter the total money obtained over 30 days from the `Normal Allowance`.

   f. Label column `C` `Doubling Allowance`. In cells `C2` through `C31` enter the amount you receive each day under the terms of the `Doubling Allowance`.

   g. In cell `C32` enter the total money earned over 30 days from the `Doubling Allowance`.

   h. Make a scatter plot with continuous lines that shows `Allowance` versus `Day`. Include both types of allowance in your graph. Make sure to label the graph correctly.

   i. Format your data so that a dollar sign is displayed to indicate currency ($0.01 instead of 0.01).

Appendix B

Extract from Galileo's *Two New Sciences*

The following quotation is from Galileo's final book, *Dialogues on Two New Sciences* [213] (179) translated by Henry Crew and Alfonso de Salvo (Macmillan, New York, 1914). Salvatore, a character who stands in for Galileo himself, has just been challenged to justify the claim that falling objects accelerate uniformly.

SALV. The request which you, as a man of science, make, is a very reasonable one; for this is the custom – and properly so – in those sciences where mathematical demonstrations are applied to natural phenomena, as is seen in the case of perspective, astronomy, mechanics, music, and others where the principles, once established by well-chosen experiments, become the foundations of the entire superstructure.[1] I hope therefore it will not appear to be a waste of time if we discuss at considerable length this first and most fundamental question[2] upon which hinge numerous consequences of which we have in this book only a small number, placed there by the Author, who has done so much to open a pathway hitherto closed to minds of speculative turn. So far as experiments go they have not been neglected by the Author;[3] and often, in his company, I have attempted in the following manner to assure myself that the acceleration actually experienced by falling bodies is that above described.

A piece of wooden moulding or scantling, about 12 cubits[4] long, half a cubit wide, and three finger-breadths thick, was taken; on its edge was cut a channel a little more than one finger in breadth; having made this groove very straight, smooth, and polished, and having lined it with parchment, also as smooth and polished as possible, we rolled along it a hard, smooth, and very round bronze ball. Having placed this board in a sloping position, by lifting one end some one or two cubits above the other, we rolled the ball, as I was just saying, along the channel, noting, in a manner presently to be described, the time required to make the descent. We repeated this experiment more than once in order to measure the time with an accuracy such that the deviation between two observations never exceeded one-tenth of a pulse-beat. Having performed this operation and having assured ourselves

---

[1] Sentences in the sixteenth century were longer than is customary in today's textbooks!
[2] The question is how falling bodies accelerate.
[3] The Author is Galileo too, of course. Salvatore is a fictional character who says whatever Galileo wants to express, but Galileo wants to make quite sure that the reader will give Galileo full credit for inventing the following experiments.
[4] 1 cubit = 1.5 feet = 45.72 cm. So the apparatus is 18 feet or 5.5 m long, and 9 inches = 23 cm wide.

of its reliability, we now rolled the ball only one-quarter the length of the channel; and having measured the time of its descent, we found it precisely one-half of the former. Next we tried other distances, comparing the time for the whole length with that for the half, or with that for two-thirds, or three-fourths, or indeed for any fraction; in such experiments, repeated a full hundred times,[5] we always found that the spaces traversed were to each other as the squares of the times,[6] and this was true for all inclinations of the plane, i.e., of the channel, along which we rolled the ball. We also observed that the times of descent, for various inclinations of the plane, bore to one another precisely that ratio which, as we shall see later, the Author had predicted and demonstrated for them.

For the measurement of time, we employed a large vessel of water placed in an elevated position; to the bottom of this vessel was soldered a pipe of small diameter giving a thin jet of water, which we collected in a small glass during the time of each descent, whether for the whole length of the channel or for a part of its length; the water thus collected was weighed, after each descent, on a very accurate balance; the differences and ratios of these weights gave us the differences and ratios of the times, and this with such accuracy that although the operation was repeated many, many times, there was no appreciable discrepancy in the results.

SIMP. I would like to have been present at these experiments; but feeling confidence in the care with which you performed them, and in the fidelity with which you relate them, I am satisfied and accept them as true and valid.[7]

SALV. Then we can proceed without discussion.[8]

Ingredients of this experiment are

1. **Precise instruments** Galileo had to invent a new kind of clock to carry out these experiments. Many experiments today continue to rely upon advances in scientific instrumentation.

2. **Controlled environment** Galileo performed the experiment under precisely controlled conditions. The channel for the ball is made as smooth as possible. The ball is as round as can be made. Channel and ball are kept clean. In short, part of experimental design is trying to manipulate the world so that it imitates mathematical perfection. There is an important preconception lying behind this effort. Physical science presumes that the real external world does have mathematical perfection, and if we only control it sufficiently, the perfection will be revealed. Galileo was challenged on this point. If his experiments only worked in a carefully controlled environment, how could one say they describe the real world? Today our point of view has changed. The laws we see operating in carefully controlled experimental environments are scientific reality, and

---

[5] The experiments were thus quite time consuming. Galileo sensed intuitively that experiments had to be repeated in order to be persuasive, and he chose quite a large number of repetitions. He did not have the advantage of modern statistics in order to explain when repeating the experiment provided diminishing returns of new information.

[6] The distance the ball travels is proportional to the square of the time the ball takes to travel it. Galileo was not able to use modern mathematical symbols as they were not yet conventional. Today we would say $d \propto t^2$ or $d = at^2$ where d is distance, t is time, and a is the constant called acceleration.

[7] It's usually only this easy to convince listeners when they are fictional and their lines are written by the Author.

[8] If only it were always this easy!

the deviations we observe from the perfect laws are due to error or chance. The goal of a good experiment is to make error as small as possible.

3. **Control of limited set of variables** Galileo's experiments involved changing two different control variables. First, with the angle of the ramp fixed, he changed the length of the ramp over which the ball rolled and measured the time needed. From these measurements he obtained the law $d = at^2$. Next, he varied the angle of the ramp and repeated the experiment, obtaining a law of the same form. He does not mention it, but the constant a must have changed each time he changed the angle of the ramp. In order to calculate the dependence of a upon the angle of the ramp, one needs trigonometry, which had not yet been developed; this may be why Galileo does not talk about the value of the constant of acceleration in these experiments.

4. **Repetition** Galileo repeated the entire experiment, he says, 100 times. This made him confident that the results were completely reproducible given the accuracy with which he could measure. Deciding what to vary and what to leave constant is one of the most difficult parts of designing an experiment. Vary too little and no laws or lessons emerge, just a single observation over and over. Vary too much, and again no laws or lessons emerge because it is impossible to distinguish random variation from the effects of the variables one controls. It is often helpful to think forward to a skeptical listener who will ask you questions about your results. Questions you are likely to face include

   • What laws or relations have you uncovered?
   • If you simply measured some numbers, why are they significant?
   • Anyone would imagine that the following things might have affected your results [temperature in the room, humidity, last Thursday's earthquake, the stomach flu epidemic, etc.] How do you know your results are not due to those effects, rather than what you claim?
   • Have you measured everything your techniques make possible? Have you developed new techniques?

5. **Precise description** Galileo's description is clear enough that the experiment can be reproduced today, and it is designed to persuade the reader of the truth of everything he claims. He may have been worried that he was establishing a standard of evidence in which the reader would be compelled to repeat the experiments before believing the results. Such a rigorous standard would have been disastrous for science, since it would have prevented science from growing through the cumulative efforts of many individuals cautiously trusting each others' results.

Appendix C

Safety in the laboratory

C.1 Safety contract

This is an overview of ways to reduce the risk of an accident and to minimize the impact of one if it does occur. The most important part of laboratory safety is the prevention of accidents. However, it is also important to understand how to respond to the various kinds of accidents that can occur in a laboratory setting. You are responsible for reviewing the following rules and following them. The most common laboratory safety equipment is listed below and you are responsible for knowing its location in your laboratory and knowing when and how to use it.

Proper attire Proper laboratory attire includes long pants, close-toed shoes, and shirts that cover the full torso. If you work with any chemicals in the laboratory, you must arrive wearing the appropriate clothing. If not, you may be asked to change clothing before engaging in your experiment, especially if you are using chemicals. Always stand – do not sit – when working with chemicals since in the event of a spill, ability to move out of the way is critical.

Cleanliness and safeness of work area You will always be expected to maintain a clean and uncluttered work area. Be sure to clean and return all materials to the appropriate storage space before leaving the lab. If left unclean and/or cluttered, you will be required to return to the lab on your own time and clean up what you left behind.

Handling and disposal of chemicals Contact your instructor for the procedures to follow when you need to handle and dispose of chemicals. **Never** simply dispose of chemicals down a drain. **Never** obtain chemicals for yourself without instructor approval and without understanding the safety considerations they involve. See Section C.2 for a flow chart to follow in deciding how to treat chemical waste.

Prohibited items in the laboratory   Smoking is always prohibited. Food and drink, including water bottles, are not allowed in the laboratory. If you are found to have food or drink in your possession, you will be asked to set them on the table at the front of the lab. After washing your hands, you will be free to get a drink or bite to eat from your stash at any time.

Location and use of safety equipment   You are each responsible for knowing where the safety equipment is located in the laboratory. Your knowledge will help your lab mates during an emergency. Handling of chemicals presents special safety challenges, and much of the equipment listed below is needed to make it safe to deal with chemicals. If this equipment is not available, you should not perform experiments involving volatile or toxic chemicals.

Fume hood: Hoods are used whenever any volatile or toxic chemical is handled. The glass window can be adjusted to stay open at several levels. Leave enough room for your hands to fit into the hood and do your work comfortably **but** make sure that the glass window is pulled down well below your eyes and nose. The fume hood fan should always be turned in the **on** position before the chemical bottle is opened. **Never** remove anything containing fuming chemicals from the hood.

Safety showers: Safety showers must be available whenever volatile or toxic chemicals are handled. Use of these showers is **strictly** for emergencies **only**! The use of these showers will result in a flood that can cause considerable damage; therefore, unnecessary use is strictly prohibited. If a chemical spills onto your clothing, you should notify your instructors immediately. They will know the appropriate action to be taken and the amount of time needed to rinse off the chemicals until help arrives.

Fire extinguishers: There is a fire extinguisher located in the laboratory. Covering a fire so that it is deprived of oxygen and extinguishes is preferable to using the extinguisher. If you see a fire, **immediately** notify your instructors and they will take the appropriate action. If, for some reason, your instructor is not around, you will need to know where this important piece of equipment is and how to use it! Take a moment to read the directions on the side of the fire extinguisher. Note what kinds of fires it can put out and how to hold it when in use.

Fire blanket: Locate the fine blanket. Remember "Stop, Drop, and Roll" if you are on fire. If you are assisting someone on fire, cover the person as completely as possible in the fire blanket to smother the flames.

Spill kit: If **any** chemical spill occurs (large or small) within the laboratory, you should notify your lab instructors immediately. They have access to a spill kit that can contain and neutralize small spills. In some larger spill cases, however, it may be necessary to evacuate the room.

Eyewash fountain: Eyewash fountains must be available whenever volatile or toxic chemicals are handled. The faucet to the eyewash can be pulled out for easier access to the hurt individual. If you get chemicals into your eyes, **immediately** notify someone in your lab and ask them to help you get to the eyewash fountain. You should make sure that either you or someone around you holds your eyelids open to the flow of water for a **minimum** of 15 minutes. This is why it is really important **not** to wear contact lenses while working with chemicals in the laboratory. Trying to get contacts out without chemicals in your eyes is hard enough – but imagine with chemicals **in** your eyes. If know you will be working with chemicals in the lab on any particular day and you have a pair of glasses, bring them to the lab to use.

First aid Accidents are most likely to occur when you are tired, on edge, or just busy with your mind on something else. The best way to prevent an accident is to observe all rules and recommendations, wear protective equipment, and stay alert for unsafe conditions. Despite all precautions and warnings, students will often find themselves involved in small accidents during their time in the lab. Each lab is equipped with a small first aid kit for accidents such as minor scrapes, cuts, burns, etc. However, no matter how big or small, it is of utmost importance to notify your instructors **immediately** in the event of **any** accident. The instructors will know what is needed, whether it be a band-aid or a call to EMS. Informing your instructor protects both you and those those around you.

I have read and understand the requirements for laboratory safety.

_____   _____

(Student Name)   (Date)

C.2 Chemical safety forms

For each chemical you use, fill out the following form.

Chemical name	Chemical formula	Handling conditions (goggles, gloves, fume hood or flame precautions)	Disposal conditions (refer to Figure C.1 for proper container)

For each chemical you use, place one label on each container used for short term storage of chemical solutions or undiluted chemicals.

Student name:	
TA name:	
Chemical name:	
Chemical formula:	
Concentration	Initial volume/mass Disposal volume/mass
Date prepared:	Disposal date:

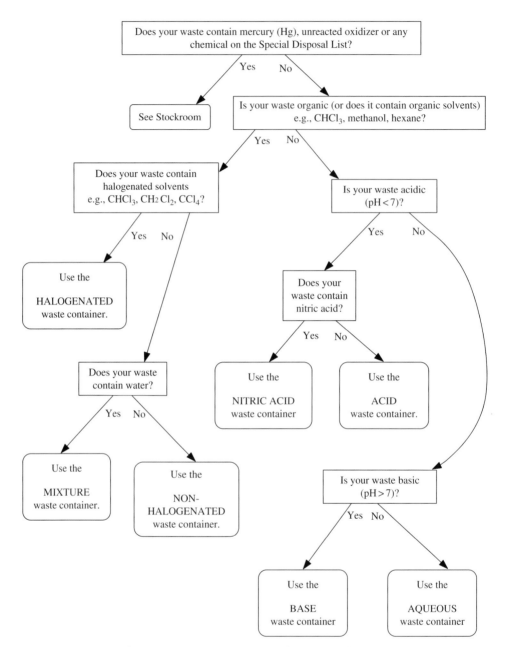

Figure C.1 Flow chart for chemical waste disposal.

Live materials, microbiological stocks, media, chemicals or supplies order form

_____   _____

(Student name)   (Date)

_____   _____

(Day and time of lab session)   (Instructor/teaching assistant)

- I need a microbiological stock of _____
  (Name of organism)

- I need _____ petri dishes with _____
        (# plates)   (Type of media)

- I need _____ml of _____ liquid culture
        (Amount)   (Type of media)

- I need other live materials:

  _____

  **(Description of live materials, including recommended supplier,
  quantitites, and catalog number)**

- I need non-living supplies:

  _____

  _____

  _____

  **(Description of non-living materials, including recommended supplier,
  quantitites, and catalog number)**

Date needed by (please allow at least one week): _____

Student cell phone:_____

Student email: _____

I understand how to properly handle, transport, and dispose of these materials.

_____   _____

(Student signature)   (Instructor signature)

Appendix D

Grading rubrics

These grading rubrics provide an attempt to formalize the way that scientists evaluate scientific work. Most of the rubric is *subtractive*. This means that the rubric describes various sorts of mistakes, and the numbers of points that can be deducted for them. The total number of points that can be deducted from each category adds to much more than 100 because scientific work can be rendered invalid by poor performance in any of these areas. Unsafe practices or plagiarism lead immediately to a failing grade.

At the very end of the rubric, there is room for additions of points. Adding points accounts for exceptional or innovative work. This part of the rubric captures the idea that the best scientific work has an element of creativity and effort that cannot be captured by the simple avoidance of doing anything wrong.

To use the rubric, the instructor will decide on the baseline value from which points will be subtracted for errors and omissions. This value may be less than 100. Getting the maximum score of 100 may require the addition of points from the final section of the rubric.

Please note that no rubric or checklist can fully capture the full range of strengths and difficulties that describe independent inquiries. This rubric provides a guide to help you identify common misconceptions and errors, but cannot fully cover all cases.

D.1 Written presentation

Student:_____ Evaluator:_____

Baseline points □	
Deduct up to	**Deduct nothing from categories that are not applicable**

Safety and motivation

Safety and ethics

of 100

Inquiry is conducted in accord with safe laboratory practice, and treats human subjects in an ethical manner

100   Inquiry has been conducted in an unsafe manner or involves illegal or unethical elements. Animals have been mistreated. Safety of human participants has been jeopardized. Example: student wants to investigate whether girlfriend's energy level varies as he varies the dosage of her prescription medication.

10   Inquiry appears to have been conducted in safe and appropriate manner, but documentation is missing or some questions about safe practice are not addressed.

5   Inquiry appears to have been conducted in safe and appropriate manner, but documentation is incomplete and some questions about safe practice are not fully resolved.

0   Student is aware of all issues surrounding ethics and safety of experiment and addresses them fully in written report. Animals and humans are treated in safe manner, following relevant guidelines. Student has obtained all necessary release forms, and completed all necessary safety training.

Motivation and time

of 20

Student is engaged by question and progress could reasonably be expected in time allotted

Deduct up to	**Deduct nothing from categories not applicable**

20	Student appears to have put no effort into project and displays no interest in outcome. Or student has settled upon project that could not possibly have been completed in limited time available. Example: in four-week project, student wants to study how size of tomatoes grown from seeds depends upon temperature. Or student spends far too little time on project. Example: student has six weeks to complete project and puts a total of one day's work into it.
10	Student appears to have put marginal effort into project and displays marginal interest in outcome, or expectations concerning time required for project were very unrealistic, or student spends too little time on project.
5	Student has put effort into project but does not appear to be interested in outcome; or, project has suffered because of poor use of time.
0	Student has worked very hard on project, displays great interest in outcome, and has used time very well.

Experimental design

of 20   *Calibration*

Measuring instrument has been configured to take meaningful data

20	Instrument provides meaningless numbers due to improper usage or lack of calibration. Example: sensors that only work when placed vertically are used in horizontal position. Example: objects weighing more than 500 gm are placed on scale giving 500 gm as maximum reading and all claimed to weigh the same. Example: 400 Hz vibrations are measured with sensor taking data at 10 Hz.
10	Accuracy of experiment is greatly reduced because of insufficient attention to calibration, range, and precision of instrument.
5	Instrument calibrated correctly on some occasions but not on others.
2	Some results suffer reduced accuracy because of lack of attention to calibration and range.
0	Instrument is correctly employed and calibrated at all times.

Deduct up to	**Deduct nothing from categories not applicable**

<table>
<tr><td>□ _{of} 20</td><td>*Error*</td><td>***Possible sources of random and systematic error have been identified, and actions have been taken to reduce them when possible***</td></tr>
<tr><td></td><td>**20**</td><td>One or more factors that have not been considered properly invalidate results. Example: measurement of the time needed for a ball to fall is dominated by human reaction time with a stopwatch. Example: student conducts a survey on whether people prefer to walk or drive by standing on a sidewalk and surveying those who walk by.</td></tr>
<tr><td></td><td>**10**</td><td>Some results invalid or very inaccurate because of factors or biases not considered. Example: a wooden block sliding down a wooden ramp is analyzed without taking friction into account. Example: survey of how typical college students feel about math courses is conducted by surveying people in physics/math/astronomy building.</td></tr>
<tr><td></td><td>**5**</td><td>Some results are not accurate or trustworthy because of factors not taken into account. Example: A measurement of the weight of air in a balloon does not compensate for weight of rubber balloon itself. Example: weight measurements of small object on precise scale are taken quickly while other students bang closet doors near scale.</td></tr>
<tr><td></td><td>**2**</td><td>Some results are questionable because of factors not taken into account</td></tr>
<tr><td></td><td>**0**</td><td>Possible sources of systematic error are understood, and have been removed to the extent possible.</td></tr>
<tr><td>□ _{of} 10</td><td></td><td>***Effects of random error are reduced to acceptable level through appropriate number of samples***</td></tr>
</table>

Deduct up to		**Deduct nothing from categories not applicable**
	10	Number of measurements is insufficient to support desired conclusions, and there is no evidence that this difficulty is appreciated. Comment: in the first stages of an experiment, there is usually little choice but to conduct a small but arbitrary number of measurements to get one's bearings. The issue is whether one learns well from this experience in the later phases of the experiment. Example: survey of whether college students want to build new student union is conducted by polling five people. Example: effect of vinegar on plant growth is conducted with two plants; one with some vinegar in soil, other without.
	5	Number of measurements hampers ability of student to arrive at conclusions. Example: effect of vinegar on plant growth is conducted with five plants with different levels of vinegar, but no two plants with the same amount of vinegar, providing no estimate of the natural fluctuations in height from one plant to another. Example: student surveys 30 people (intrinsic uncertainty on order of 20%) trying to learn about presidential race known to be quite close.
	2	Number of measurements is adequate to support desired conclusions, but appears to have been arrived at in completely arbitrary way. Example: to see whether a scale gives reproducible results, student weighs penny on scale 100 times, even though the measurement never differs by even one part in 1000 for any two trials.
	0	The formula $\Delta x = s/\sqrt{N}$ or other appropriate equation is used to calculate number of measurements to reach desired accuracy once preliminary data have been obtained, and appropriate quantities of data have been obtained.
☐ of 10 *Variables*		***Experiment is carefully designed to vary control, measure response, and keep other variables constant***

Deduct up to	**Deduct nothing from categories not applicable**

10   Experiment is invalid because of failure to identify clearly quantities held fixed and quantities allowed to vary. Important variables vary from trial to trial and are not even recorded. Example: student wants to know how growth of plant depends upon sunlight, but keeps plant outside and does not record temperature or rainfall.

5   Validity of experiment is somewhat reduced because of failure to identify clearly quantities held fixed and quantities allowed to vary. Example: student wants to see how manageability of hair depends upon meals eaten, but eats different foods each day the experiment is being conducted. Example: student records number of people walking past his house between 9 and 10 each morning on 10 consecutive days, but without recording anything about weather or even day of the week.

2   Quantities to be varied and quantities to be held fixed are clearly identified, but the number of quantities being varied is too great for a persuasive experiment to be possible in the time allotted.

0   Student has decided to vary a limited number of quantities, and measure a limited number of quantities, and has thought carefully about how to keep everything else constant.

Analysis

|of 10|

Graphs   ***Displays data in graphical form, including histograms or functional relations as appropriate***

10   Large quantities of data have been taken, and are either reported in large tables of data or not at all, and no graphs are presented.

5   Some important data that should have been represented in graphical form have not been.

2   Graphs have been prepared when necessary, but some are confusing or improperly prepared.

0   Makes use of appropriately constructed graphs whenever needed in order to explain results.

Deduct up to		**Deduct nothing from categories not applicable**

<table>
<tr><td>of 5</td><td></td><td>***Indicates error bars on measurements as appropriate***</td></tr>
<tr><td></td><td>5</td><td>There were repeated trials for various measurements, but no error bars have been computed or recorded, or else they have been computed incorrectly.</td></tr>
<tr><td></td><td>0</td><td>The standard error has been computed for all repeated trials and recorded on plots as error bars.</td></tr>
<tr><td>of 5</td><td></td><td>***All axes are labeled and units clearly indicated***</td></tr>
<tr><td></td><td>5</td><td>Axes are not labeled.</td></tr>
<tr><td></td><td>0</td><td>Axes are labeled.</td></tr>
<tr><td>of 5</td><td></td><td>***Figure is numbered, has a caption, and is referred to in the text***</td></tr>
<tr><td></td><td>5</td><td>Figures are not numbered, lack captions, or are not referred to in the text.</td></tr>
<tr><td></td><td>2</td><td>Some ingredients missing, such as some figures without numbers.</td></tr>
<tr><td></td><td>0</td><td>Figures are numbered, have a caption, and are referred to in the text.</td></tr>
<tr><td>of 5   *Statistics*</td><td></td><td>***Computes means and standard errors for measured variables***</td></tr>
<tr><td></td><td>5</td><td>Performs repeated trials, but does not compute the average or other measure of central tendency, or interprets them incorrectly.</td></tr>
<tr><td></td><td>0</td><td>Calculates, reports, and correctly interprets means and standard errors.</td></tr>
<tr><td>of 20</td><td></td><td>***Makes appropriate use of statistics***</td></tr>
<tr><td></td><td>20</td><td>Has neither null hypothesis nor testable question, and arrives at conclusions without any statistical arguments.</td></tr>
<tr><td></td><td>10</td><td>Has testable question, but fails to use statistics when interpreting data: Example: calculates means of different quantities, and concludes that one is larger than the other without considering whether the difference could be due purely to chance.</td></tr>
</table>

Deduct up to	**Deduct nothing from categories not applicable**

5   Attempts to use statistics to understand significance of outcomes, but makes technical errors. Example: uses χ^2 test appropriate for integer samples on continuous data.

0   Employs statistics to establish significance of results. Example: uses the standard error associated with means to decide whether the difference between means is significant. In a first pass, this could be performed by looking at graphs, and next if needed by use of Z or t tests. In experiments with complex design, a more sophisticated test such as ANOVA may be appropriate.

of 20	*Modeling*	***Constructs simple predictive relations, ranging from fits with simple functions to recursion relations or differential equations, analyzes their consequences in programs such as Excel, and compares with data***

20   Experiment involves quantities that vary in regular fashion in time or space, but student makes no attempt to construct predictive model, not even one employing just elementary functions. Level of predictive modeling that can be expected depends upon mathematical sophistication. Many apparently simple situations, such as objects acted upon by friction, may require use of calculus.

10   Constructs predictive model that is completely inappropriate. Example: blindly clicks on a randomly chosen regression function in Excel and presents the formula. Example: uses addition in a formula where multiplication is appropriate.

5   Student makes some technical errors in construction of predictive relation. Example: presents adequate linear fit in case where data are clearly fit by exponential function.

0   Constructs correct predictive relations, ranging from simple formulas that might be based upon dimensional analysis, up to and including recursion relations or differential equations, analyzes their consequences in programs such as Excel, and compares with data. Some classic patterns that should be familiar to almost everyone include exponential growth and decay in their many manifestations.

Deduct up to	**Deduct nothing from categories not applicable**

Sources

of 10	*Literature*	***Makes use of research literature to answer questions outside scope of project as needed***

10 Asks questions that cannot possibly be answered by isolated inquiry, and makes no attempt to learn any background from books or articles. Performs no searches in case where they are needed and appropriate

5 Employs sources to provide background to inquiry, but chooses inappropriate or unreliable sources. Example: relies exclusively on rapid search with Google or other general search engine and obtains inappropriate materials. Articles that have passed through peer review are generally much more reliable than websites. Many websites contain information from people without special knowledge, that have not been checked by those who have, and cannot be trusted.

0 Makes use of books and articles as inquiry progresses to deepen understanding. Even in research areas that are highly developed, there can be times when it is extremely profitable for researchers to try to think matters through completely on their own without consulting sources. Different people have different styles. However, if experimental data are needed for an inquiry, and if gathering them is beyond one's resources, then one really must look them up.

Written presentation

of 20	*Overall quality*	***Clearly written with correct grammar and spelling***

20 Inquiry report is carelessly written in very poor English, and is impossible to understand.

10 Inquiry report is carelessly written in very poor English, and is difficult to understand.

5 Inquiry report is written in poor English that detracts from its quality.

2 Inquiry report is by and large well written, but contains occasional errors in grammar or spelling that detract from its quality.

0 Inquiry report is well written in clear, grammatically correct, and engaging English.

Deduct up to		**Deduct nothing from categories not applicable**
☐ of 10	*Abstract*	***Clearly explains purpose of project and summarizes main conclusions***
	10	Report does not contain an abstract.
	5	Abstract is taken verbatim from proposal and is written in future tense without any description of results.
	0	Abstract allows a broad audience to learn main conclusions from project by reading a few sentences.
☐ of 10	*Design*	***Design of experiment or survey is clearly explained with appropriate mixture of words and diagrams***
	10	There is no description of design or experimental procedure. In surveys, there is no description of large population sample it is supposed to represent.
	5	The procedure used in the experiment is described but is difficult to understand and would be hard to reproduce. In surveys, description of the large population the sample should represent is vague and not completely convincing.
	0	Report contains a detailed description of the procedure, providing enough information that someone else could reproduce the experiment, and also providing guidance about attempts that did not work, and explaining why the procedure employed was successful. For surveys, report carefully describes large population the sample is supposed to represent, and explains steps taken to make the sample representative of the whole.
☐ of 100	*Literature*	***Provides citation for all sources employed in study and explains their significance***
	100	Plagiarizes report.
	20	Uses unattributed material from internet or other sources.
	10	Makes use of background sources and places material in own words, but without attribution. Makes no use of peer-reviewed sources at all, although they would have been appropriate.
	2	Makes use of inappropriate sources, although with correct attribution.

Deduct up to		**Deduct nothing from categories not applicable**
	0	Uses appropriate sources as needed; the default expectation is for three peer-reviewed sources by the final inquiry. Uses quotations up to 50 words, with attribution. Puts longer passages into own words, again providing citation.

of 10

Engages in critical analysis of sources

10 In areas where controversy still exists, simply reports what one side or the other says with no discussion. Example: a phenomenon called "cold fusion" was announced to the media in the mid-1980s. The phenomena could not be reproduced in other laboratories, and most articles in peer–reviewed journals say they do not exist. Some of the scientists who found the phenomenon started their own journals so they could publish, but these have no credibility. By contrast, "high-temperature superconductivity," which was found a bit earlier, was rapidly reproduced at many laboratories, and has been a standard scientific topic ever since.

5 Acknowledges differing points of view within the scientific community, but overlooks or misrepresents some important points of view.

0 Is aware of and acknowledges differing points of view within the scientific community, and uses reasoned arguments to explain why he or she has adopted a particular view. Example: most environmental scientists believe that the planet is warming, and that steps should be taken to slow the production of greenhouse gases. However, the position is not universally held, and it would be legitimate for a student to explain why he doubts the majority view, and to perform an inquiry to check claims that are customarily made

of 10 *Collaborators*

Acknowledges all collaborators on project and explains their role

10 Does not mention collaborators in written report.

5 Mentions name of collaborator, but not role in project

0 Acknowledges all collaborators on project and explains their role

Deduct up to		**Deduct nothing from categories not applicable**

of 10	*Analysis*	***Thoughts leading to analysis of data are clearly explained, with all graphs and computations described in words***
	10	Graphs or equations are presented with no explanation.
	5	Graphs or equations are presented with confusing or incomplete explanations.
	0	Clear verbal explanations accompany all graphs and equations, explaining their significance and how they were obtained.

of 10	*Conclusions*	***Consequences of study are clearly communicated, neither generalizing too much, nor omitting important findings***
	10	Report terminates without conclusions.
	5	Concludes with statements that do not honestly reflect conclusions that could be reached from work described in the body of the presentation. Sometimes, these may represent wishful thinking, or a desire to impress the teacher.
	0	Thoughtfully summarizes the main results, acknowledging both strengths and weaknesses of what has been accomplished, and indicating new things that might be interesting in the future.

	Additions	***Student performs exceptional work***
	+20	Student obtains a research result that might be publishable.
	+10	Student has put enormous effort into project, or student has come up with an unusually innovative method for solving a problem given the time involved, or report is exceptionally well written.
	+5	Student has performed an excellent job of looking into background material and finding references.

of 100	***Total written***	

D.2 Oral presentation

Student:_____   Evaluator:_____

Deduct up to	**Deduct nothing from categories that are not applicable**

	Presentation style		

<table>
<tr><td>of 50</td><td>Presentation style</td><td colspan="2">Speaks clearly and at audible volume, engages audience, employs visible font sizes, and uses graphical materials or live demonstrations whenever possible to illustrate ideas</td></tr>
<tr><td></td><td></td><td>50</td><td>Presentation is impossible for audience to understand because of poor use of spoken language and visual tools.</td></tr>
<tr><td></td><td></td><td>20</td><td>Presentation is very difficult for audience to understand because of poor use of spoken and visual tools.</td></tr>
<tr><td></td><td></td><td>10</td><td>Presentation is understandable, but either the written words and graphics or spoken presentation need improvement. Example: many slides use fonts too small for many in the audience to see. Example: talk is presented too softly for many in the audience to hear, or throughout seems tentative and uncertain.</td></tr>
<tr><td></td><td></td><td>5</td><td>Presentation is effective, but could be improved in places by more effective speaking style, or better use of written words and graphics. Example: some slides have written material written in small fonts that are difficult for some to see. Example: speaker does not always speak loud enough for whole audience easily to hear. Example: everything speaker says is written on slides, and speaker simply reads slides to audience.</td></tr>
<tr><td></td><td></td><td>0</td><td>Presentation is easy for all audience members to hear and understand, it is engaging, and the presentation makes effective use of spoken words and visual materials to convey understanding.</td></tr>
<tr><td>of 10</td><td>Introduction</td><td colspan="2">Describes motivation for project</td></tr>
<tr><td></td><td></td><td>10</td><td>There is no introduction to the project. Plunges into details of project with no attempt to engage audience.</td></tr>
<tr><td></td><td></td><td>5</td><td>Explains why project is interesting, and attempts to involve audience, but introduction needs substantial improvement. Example: speaker refers to advanced concepts in biology, not recognizing that half the audience consists of math majors.</td></tr>
</table>

Deduct up to			**Deduct nothing from categories not applicable**
		0	Clearly explains why project is interesting, and involves audience.
of 10	*Background*		***Provides clear summary of relevant background knowledge***
		10	Does not describe any relevant background knowledge
		5	Uses technical terms unknown to the audience with no explanation, and refers to theories and ideas that are never explained.
		0	Given time constraints, uses words and pictures to teach as much relevant background as feasible. Defines all technical words that are used later.
of 10	*Design*		***Describes project design, explaining major choices, using diagrams and illustrations***
		10	Does not describe project design.
		5	Makes vague reference to procedure, without explaining to anyone how goals were reached.
		0	Clearly describes experimental procedure, making good use of words and images.
of 10	*Results and data*		***Presents main results***
		10	Does not present main results.
		5	Presents main results, but they are very difficult to understand. Example: presents screens full of unreadable charts full of numbers
		0	Clearly and accurately represents main conclusions arrived at in study. Concisely conveys statistical certainty with which results were reached. In most cases, these goals are best met by presenting results in graphical form with error bars, and summarizing them well in spoken and written words.
of 10	*Conclusions*		***Summarizes findings***
		10	Comes to a halt without concluding statements.
		5	Closing points are confusing.
		0	Closes with a clear statement of the main points to remember.
of 100	*Total oral*		

Index